I0038427

Organic Agriculture Handbook

Organic Agriculture Handbook

Edited by Herbert Wright

SYRAWOOD
PUBLISHING HOUSE

New York

Published by Syrawood Publishing House,
750 Third Avenue, 9th Floor,
New York, NY 10017, USA
www.syrawoodpublishinghouse.com

Organic Agriculture Handbook
Edited by Herbert Wright

© 2023 Syrawood Publishing House

International Standard Book Number: 978-1-64740-345-4 (Hardback)

This book contains information obtained from authentic and highly regarded sources. Copyright for all individual chapters remain with the respective authors as indicated. All chapters are published with permission under the Creative Commons Attribution License or equivalent. A wide variety of references are listed. Permission and sources are indicated; for detailed attributions, please refer to the permissions page and list of contributors. Reasonable efforts have been made to publish reliable data and information, but the authors, editors and publisher cannot assume any responsibility for the validity of all materials or the consequences of their use.

Trademark Notice: Registered trademark of products or corporate names are used only for explanation and identification without intent to infringe.

Cataloging-in-publication Data

Organic agriculture handbook / edited by Herbert Wright.
    p. cm.
Includes bibliographical references and index.
ISBN 978-1-64740-345-4
1. Organic farming. 2. Agriculture. I. Wright, Herbert.
S605.5 .O74 2023
631.584--dc23

TABLE OF CONTENTS

PREFACE

Every book is a source of knowledge and this one is no exception. The idea that led to the conceptualization of this book was the fact that the world is advancing rapidly; which makes it crucial to document the progress in every field. I am aware that a lot of data is already available, yet, there is a lot more to learn. Hence, I accepted the responsibility of editing this book and contributing my knowledge to the community.

Organic agriculture is a holistic approach to agricultural production system that sustains the health of soils, ecosystems and people. It relies on ecological processes, biodiversity and cycles adapted to local conditions, rather than the use of inputs with adverse effects. Agroecology is related to the study of organic farming methods, and studies ecological processes applied to agricultural production systems. The practice of organic agriculture relies on the use of organic fertilizers such as compost manure, green manure and bone meal. Crop rotation and companion planting are major agricultural techniques employed in organic farming. It also encourages crop diversity. Planting a variety of vegetable crops is beneficial for the overall farm health as it supports a wide variety of insects and soil microorganisms that are beneficial for agricultural environment and help protect different species from becoming extinct. This book unravels the recent studies in the field of organic agriculture. It will provide comprehensive knowledge to the readers.

While editing this book, I had multiple visions for it. Then I finally narrowed down to make every chapter a sole standing text explaining a particular topic, so that they can be used independently. However, the umbrella subject sinews them into a common theme. This makes the book a unique platform of knowledge.

I would like to give the major credit of this book to the experts from every corner of the world, who took the time to share their expertise with us. Also, I owe the completion of this book to the never-ending support of my family, who supported me throughout the project.

Editor

Environmental Aspects of Organic Farming

Jan Moudrý jr. and Jan Moudrý

1. Introduction

In the nature, there is no viable ecosystem that can work without any negative feedback. Any interference with the system does not affect only in one way but it necessarily evokes another often unpredictable reactions. In Europe these days, there are quite natural self-regulating systems with closed energy flows rather exceptionally. This is due to a significant environment disturbance by humans. In terms of area, the main human activity interfering in natural ecosystems was agriculture.

Agroecosystems are tightly connected with more natural, respectively near natural ecosystems. The mutual relationship is bidirectional (nutrient, organism and energy flows, impact on microclimate). In terms of ecological stability, agroecosystems show all disadvantages of juvenile (immature) ecosystems.

Some of the agroecosystem characteristics are:

- additional external energy inputs,
- significantly reduced biodiversity,
- artificial support (selection) of dominant production species,
- juvenile stage of succession (anthropogenic disclimax),
- reduction even paralysing of self-regulatory processes,
- significantly reduced degree of environmental stability
- irreversible degradation processes occurrence

Under the European conditions, there were historically evolved the mixed, commercially oriented, permanent, mechanized systems with high energy-material inputs, i.e. intensive farming systems.

The external manifestation of the intensive agroecosystem is a high degree of landscape urbanization (natural vegetation suppression, sharp land boundaries, the amount of built-up areas, etc.). At the field level, the typical feature is the stand uniformity, inability to self-regulation, often poor ability to environmental adaptation, permanent soil erosion and the need to control other material and energy inputs. Intensive agroecosystems represent a significant spatial landscape heterogeneity reduction and the corresponding species diversity decline.

In real-life working, the intensification is achieved in many ways which are often combined. In particular, the production is narrowly specialised (the number of cultivated plant species is decreasing to monocultures, with livestock, the specialization goes down to level of individual category breeding with no ties to the land and crop production).

Significant intensification factors are concentration (production organisms density increase in time and space), step land use (multistorey stables), high degree of mechanization even technological processes automation, intensive use of additional chemical inputs, energy and information.

Highly intensive mechanized system is becoming completely dependent on external inputs (machinery, fuel, chemicals, seeds). High external energy-material inputs strongly reduce the systems energy efficiency. The ratio of energy input to energy gained from the crop is up to 3:1 while with non-intensive systems, it is 1:20 and more. Within highly intensive mechanized livestock production system, the energy balance is even less effective. However, these systems are very effective in the short term in terms of labour productivity and land utilization.

On the contrary, extensive (low input) farming systems have almost the opposite characteristics. Their main feature is the external input reduction. Extensive agroecosystems are characterized by lower energy and material flows per a unit of area and usually higher diversity, less need for external intervention and greater stability and self-regulatory abilities. They significantly contribute to the conservation of natural resources. Lower inputs can be compensated by a quality management. Reducing inputs usually brings an agroecosystem production capacity reduction. Lower yields can be realized at a lower cost without a significant profit reduction.

In the world, there are extensive farming systems on 80% of the area and on 20%, there are intensive farming systems. The general trend is the increasing agricultural production intensification in many developing countries (China, Brazil, Russia,...) and chemical inputs reduction, respectively their substitution by biological or rational means in developed countries, especially in the EU. Due to the growing human population and its demands on sufficient of varied and quality food, a certain degree of agroecosystems intensification is necessary. However, it is crucial that agroecosystems have a sustainable character.

According to the simple OECD definition, for sustainable agroecosystems, there can be considered those that meet the needs of these days and do not limit the future generation. The following definition is more precise: *"Sustainable agroecosystems-agricultural and food systems are economically viable, meet society's need for food assurance, while they retain and enhance natural resources and environmental quality for future generations."*

From the definition, it is clear that the sustainable agroecosystem does not carry only the function of food and row material production. The organic farming system is focused on the homogenization of landscape production and non-production functions, where the emphasis is laid on environmental aspects. These include in particular:

1. Maintenance and improvement of soil fertility.

2. Nutrients recirculation and a prevention of the entry of extraneous substance into agroecosystem.

3. Water management in landscape and its protection against contamination.

4. Air quality improvement and greenhouse gas emission reduction.

5. Genetic resources protection and biodiversity maintenance.

6. Preservation of landscape features and their harmonization.

7. Efficient use of energy, focusing on renewable resources.

8. Optimization of life for all organisms, including humans.

Organic farming is based on the principles of sustainable farming and therefore it is a model for the sustainable agroecosystems establishment.

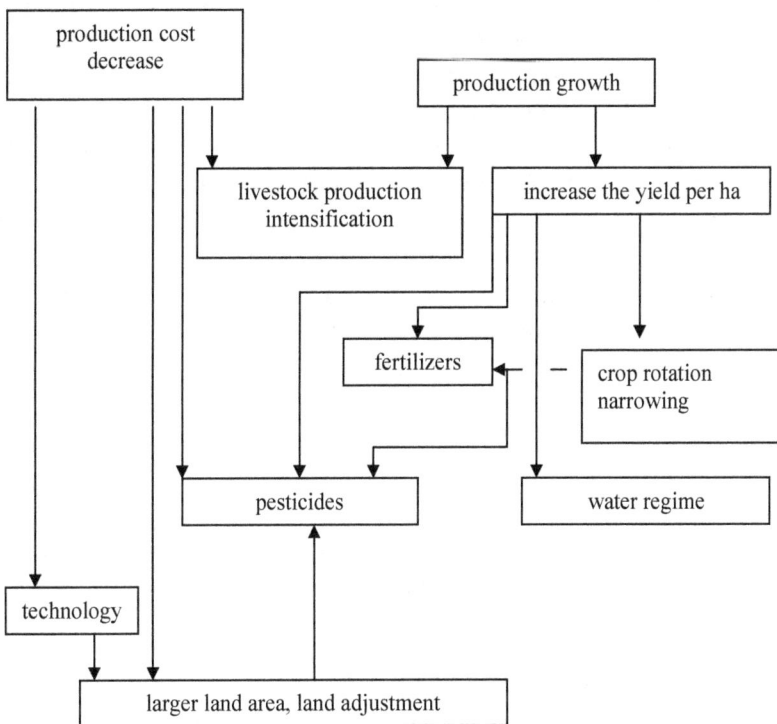

Scheme 1. Impact of market-oriented production on relationships in agroecosystem [54].

1.1. Soil environment

Soil is one of the most important natural resources and plays a key role in agriculture. Healthy soil is essential for growth and evolution of healthy plants. In addition to the production function, the soil has many other functions such as filtering, buffering, transformation and it is the environment for organisms and also its socio-economic function is not negligible.

There are following positive changes within organic farming:

a. soil organic matter (up to 30% higher organic carbon content),

b. increased soil biological activity (by 30-100 %), biomass decomposition indicator,

c. higher total edaphon biomass (by 50-80 %),

d. higher saprophytic fungi abundance, higher root colonization by mycorrhiza,

e. more efficient use of acceptable resources by soil microorganisms,

f. improved physical and chemical soil properties, soil structure,

g. improved hydroscopicity and erosion threat reduction

The soil organic matter research is mostly concentrated on the organic carbon content and its changes during conversion to organic farming. Many studies have confirmed that areas under organic cultivation have a higher organic carbon content as compared with areas under conventional cultivation. However, in some researches, there was a higher decomposition of organic matter such as within more intensive soil cultivation associated with mechanical weed control. However, long-term experiments have confirmed the hypothesis that organic farming methods better protected the soil organic matter. The research also points to a larger amount of humic substances. An important factor in the soil organic matter protection is the minimum soil cultivation. A properly designed crop structure, fertilizers, etc. are also important. A higher supply of organic matter in the form of crop residues and organic fertilizers creates favourable living conditions for soil fauna. The soil provides a habitat for a large number of various organisms. The positive role of organisms consists mainly in organic matter decomposition and inorganic substances transformation where nutrients are more accessible for plants and where is also a synthesis of complex organic substances enriching humus reserves in the soil. In organic farming, the key role for nitrogen plant nutrition is played by a symbiotic fixation with papilionaceous plants. In the soil, there is also a nonsymbiotic fixation, e.g. by in-the-soil-free-living heterotrophic aerobic bacteria. Rhizosphere is a zone where the main part of nutrient cycle takes part due to the interaction between soil, roots and microorganisms colonizing the plant root environment. In organically cultivated soils, [35] has observed by 40% more mycorrhiza than in soils within integrated farming.

Natural and active edaphon contributes to the protection of plant roots against parasite and pathogen attack but also to degradation of toxic substances which enter the soil within the chemical plant protection, environmental contamination from industry as well as metabolic products of other organisms. The soil deterioration leads to a biodiversity reduction. Biological degradation of soils is usually associated with their physical and chemical degradation.

The soil liveliness is indicated by a number of indicators: An important role is played by earthworms which are subjects of many studies precisely because of their sensitivity to the soil environment disturbance. Organic farming has up to 50% more biomass and abundance of earthworms as compared to integrated farming, greater biodiversity of earthworm species, changes in the population composition indicated by a larger number of juvenile earthworms. Earthworms are useful-they aerate and mix the soil, help with organic matter decomposition [35].

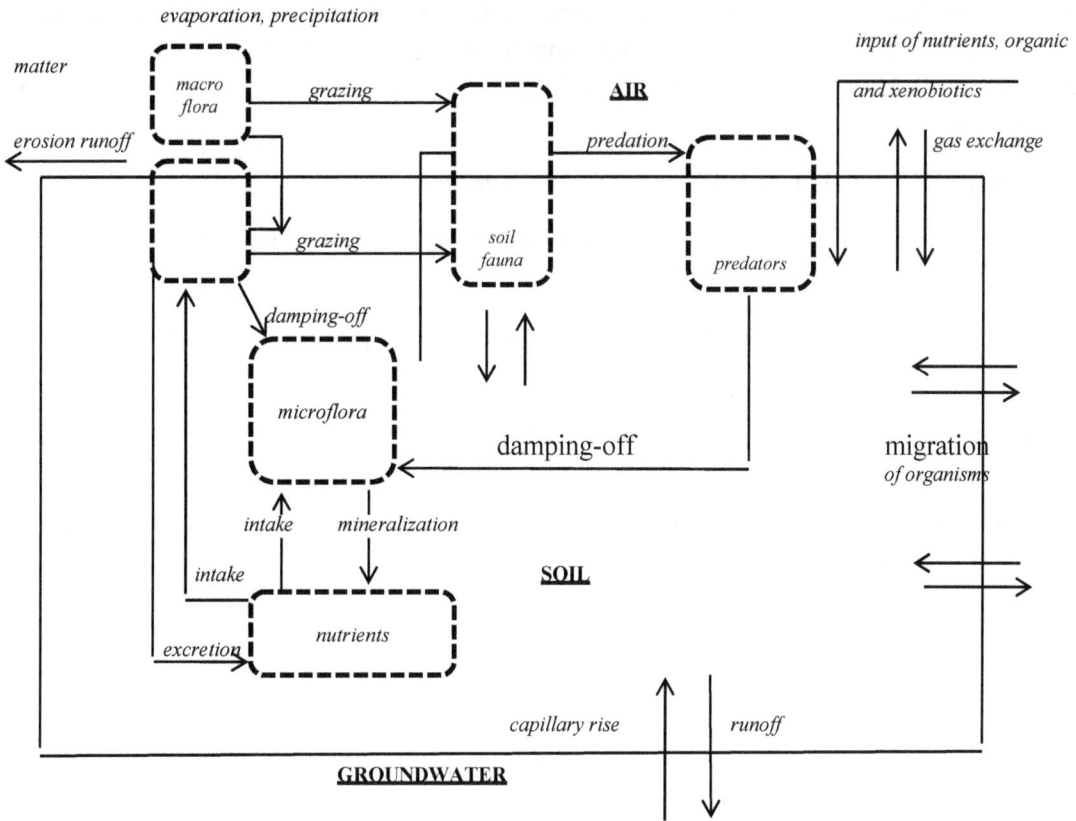

Scheme 2. Diagram of the soil ecosystem function (adapted according to [54])

An important indicator of the soil organic matter decomposition is the biological activity. However, it is possible to state that changes on biological activity are slow and in many studies comparing organic and conventional farming systems, there were no differences experienced. Some changes (species diversity increase, biomass production decrease resp. yield decrease) occur almost immediately, other (increase of natural soil fertility, soil organic matter content, system stability) show up in the longer term.

On the other hand, the soil deterioration leads to a biodiversity reduction. Biological degradation of soils is usually associated with their physical and chemical degradation. The biodiversity of soil microorganisms is reduced by an intensive cultivation through the use of mineral fertilizers and pesticides. Soils with low humus content and light soils are relatively more sensitive.

A serious problem on large areas, especially of arable land, is a water and wind erosion. Organic farming has a positive impact on its reduction thanks to the more diverse crop rotations with a higher share of clover and grass-legume mixtures, a higher percentage of catch crops and underseeding prolonging the soil cover over the year, a lower representation of wide-row crops (e.g. corn), a more intensive organic fertilization and other factors. Nevertheless, a danger of erosion can occur also in organic farming (and sometimes even more than in conventionally cultivated areas) in particular, because of more frequent mechanical tillage or slower plant development due to a lower mineral nitrogen content in the soil. Structural soils are well more resistant to erosion [50]. When comparing particular factors, we find that positives predominate.

A quality soil ecosystem should meet following criteria:

• water flowing out from the ecosystem should have such a purity that it is suitable for drinking water treatment;

• growth of crops and their composition in terms of consumption should be at an acceptable level;

• microbial processes in the soil should be natural, therefore, relationships between microbial biomass, microbial activity and soil organic matter should be predictable;

• the soil should not contain potentially toxic chemicals (organic and inorganic) in concentration that should affect the previous criteria;

• physical soil properties should enable the normal function of the ecosystem.

1.2. Nutrient recirculation

In agricultural ecosystems, there can be the soil fertility increased by additional inputs such as manure or fertilizer application. Plant nutrition within conventional farming is more dependent on the input of nutrients in the slightly soluble form, predominantly from synthetic fertilizers. A part of nutrients leaves the system as a loss. On the contrary, the natural ecosystem fertility depends almost entirely on natural biological processes, i.e. nitrogen fixation and soil organic matter mineralization. In organic farming, there is the concept of soil as a living system. Therefore, the fertilization system is designed in order to respect natural nutrient cycle and not to adversely affect complex biological processes which nutrient cycles are dependent on. Organic approaches lead to a higher organic matter content in the soil while conventional intensive farming on the arable land can lead to a reduction of this matter content. Assigning of legumes and clover into crop rotations is of a considerable importance because of their agromeliorative effect on the soil. The effect is manifested also on physical properties with an effect on the soil bulk density, water holding capacity, increasing porosity, soil structure stability, etc. The size of invertebrate organism population depends on the physical condition of the soil. They usually require well aerated non-compacted soil with low density.

A different system of fertilization in conventional and organic farming can have both a direct and indirect effect on soil organisms. The direct effect is related to the composition and the amount of applied fertilizers and the indirect effect is connected with changes of physical and chemical soil properties.

One of the basic principles of organic farming is the most closed nutrient cycle, minimal nutrient loss and limited nutrient supply to the system. In order to maintain the soil production capacity, it is necessary to replace nutrients drawn from soil by harvests and lost nutrients by biologically transformed organic matter in the soil. Regular supply of organic matter into the soil is ensured by crops grown in order to enrich the soil with organic matter (clover, catch crops for green manure), crop residues, residues of cultivated plant roots and manure. Soil organic matter serves as a continuous reservoir of nutrients and energy for the soil environment. It is also a factor of soil environment stabilization. Soil organic matter is a source of nutrients for grown plants, source of energy for soil microorganisms, improves physical and chemical soil properties, water regime, increases decontaminating and buffering soil capacity and decreases nutrient losses washed away from soil, increases antiphytopathogenic soil potential and strengthens the plant immune system.

Growing crops without mineral fertilizers is possible under following assumptions:

- consistent application of all manure

- direct application of recycled biomass and by-products into the soil

- compost production in compliance with the technological process

- use of uncontaminated nutrient resources

- use of peat and humic substances in order to improve habitat and nutritional status of crops

- use of indirect fertilizers containing nitrogen fixating bacteria (free-living in the soil, rhizobia) or bacteria that access e.g. S, P, K and other nutrients form soil reserves.

The function of soil organisms is for transformation of organic matter in the soil irreplaceable. In the initial stage of organic matter transformation, zooedaphon participates in the destruction of crop residues and in the production of organomineral components. The transformation proceeds on depending on organic matter composition, edaphon activity and environmental conditions in terms of mineralization and humification. For the humus formation, there are important root exudates, roots of perennial fodder plants, legumes, manure and compost and dead residues of zooedaphon. A ready potential source of nutrients in the soil is the primary organic matter, root exudates and manure. Crop residues alone cover mineralization losses of about 50%.

Of the total amount of organic matter, less than 10% is humified. The increase of the amount of permanent humus in the soil is a matter of long-term (tens to thousands of years). Water-soluble carbohydrates decay fastest, cellulose, hemicellulose and protein decay average rapidly and lignins and pectins are the slowest. The rate of decomposition of different organic matter sources is greatly different. Sustainable land management system assumes a balanced budget of organic matter in the soil consisting in replacing mineralized organic matter by inputs in the range of mineralization and losses.

Substance	Crop		
	corn	alfalfa	wheat
Water-soluble sugars	6.72	4.36	4.68
Hemicellulose and starch	42.61	14.85	23.30
Cellulose	23.29	32.25	42.12
Proteins	4.75	16.44	4.31
Lignin rest	18.27	29.60	23.00

Table 1. Chemical composition of crop residues (% dry matter).

1.3. Water in landscape

Disposal of pesticides and morforegulators from the organic farming system reduces significantly the contamination of the environment including surface-and groundwater by residues of these substances. In the area of organically cultivated fields, there are surface water and groundwater less contaminated with plant protection products. These substances harm also to aquatic animals, even at low concentrations (below detection limit) [33].

Also the prohibition of use of slightly soluble synthetic nitrogen fertilizers in organic agroecosystems reduces significantly the load of surface-and groundwater by nitrates. Reduced animal surface load of soil (limit for organic farming 1.5 LU/ha, optimum 0.4-1.0 LU/ha in relation to site conditions), optimal manure use (method of treatment and storage, time and rate of its application, higher use of green manure and liquid manure restrictions), appropriate use of leguminous plants and atmospheric nitrogen bounded by them, appropriate crop rotation and an effort to maximize the vegetation cover (catch crops and permanent crops), soil conservation cultivation methods and erosion reduction and other impacts contribute to this. Within organic farming, there are by 35-64% less nitrates washed away as compared to conventional farming agricultural plants [53]. In 40 scientific publication comparing nitrate leaching or a leaching potential analysed by [22], twenty eight stated lower values within the organic farming system, nine issued comparable data and only in three cases, the nitrate leaching respectively its potential were higher within organic farming than in conventional one. Yet, there were two critical areas for potential water pollution recognized and studied within organic farming. These are manure composting and farming with residual nitrogen from leguminous plants. Storage and composting of manure on unpaved surfaces can cause leakage and subsequent contamination of groundwater and surface water. A significant leaching can also occur when the nitrogen source accumulated by leguminous plants is inappropriately used, i.e. by ploughing alfalfa in autumn, followed by sowing crops with low demands on soil nitrogen content [48].

Nutrients from intensively cultivated cropland load water due to overland flow. Within organic farming, this risk is reduced by greater ruggedness of landscape, more extensive integration of landscape features as well as the crop diversity and their optimal distributing. Marginal strips along water courses and reservoirs and protective grass strips on slopes make buffering areas, limit overland flow and prevent erosion and expand biodiversity. Organic farming is supposed to be the preferred farming system especially in the areas of water resources conservation.

Organic farmers fertilize the soil in such way so not to pollute groundwater. Within organic farming in addition to manure and liquid manure, there is green manure also used as fertilizer and legumes are properly incorporated into the crop rotation. This reduces the leaching of nitrogen into groundwater. As a result of leaching through the soil profile and due to erosion and surface runoff, nitrates in water causes contamination of the hydrosphere and along with phosphorus cause eutrophication [30]. Concurrently, there is a leaching of base cations (K+, Ca2+) and thus the upper soil layer are depleted of these nutrients. Indirectly, this leads to acidification. Due to leaching, several tens of kg N. ha-1. year-1 is normally lost [51]. Organically cultivated areas provide better flood protection than conventional surfaces. The high infiltration capacity of the soil with virgin structure may reduce the intensity of floods [49].

1.4. Air quality

Agricultural activities have not a negligible impact also on air quality. Organic farming as a whole contributes to the creation of anthropogenic greenhouse gases with about 14% while the ratio differentiates with particular countries according to the agricultural production intensity. Due to its large area impact, agriculture belongs to the largest producers after industry and mining. [14] state that agriculture contributes to annual increase of GHG emissions with approximately one fifth. Even higher value is stated by [10], whose findings report the proportion of 27%.However, with the increasing consumption of food and agricultural intensification, this percentage is rising. When adding pre-farming and post-farming phases to the agricultural frame itself or quantifying food life cycle, the emission load is even higher. This is mainly due to production of agrochemicals and processing of primary agricultural production. Moreover due to the increasing human population, agriculture will even increase its pressure on the environment. There is constantly running the conversion of natural habitats into agroecosystems and in parallel the intensification of farming on existing agricultural land. This is largely accompanied by other chemicalization of agriculture. The pressure on increasing yields and the food consumption grow aggravate the share of agriculture of emission load production. However in most cases, organic farming produces lower emission load, not only in the field phase but also in the consequential phases.

Agriculture produces emissions in many ways. For example, CO_2 is released during the consumption of fossil fuels or within reduction of organic matter content in the soil. N_2O is released as a result of fertilizer application and within soil processes, CH_4 from the digestive tract of some livestock species. Especially in the crop production, the emission production is influenced by the intensity and thus by the system of farming.

Organic farming has a number of tools which help to reduce emission loads (see tab. 2). [42] in accordance with the IPCC fourth assessment report states as the optimal measures for mitigation (reducing stress) in organic agriculture the following points:

- crop rotation and character of the agricultural system

- management of nutrients and fertilization

- livestock, improving the pasture utilization and fodder supply

- soil fertility management and restoration of degraded soil

Within the environmentally friendly approach, organic farming systems generally seek more precisely to work with energy and to minimize inputs and to close the farm cycle as soon as possible. This leads to a large emission reduction particularly due to the reduction of synthetic nitrogen fertilizers whose production is among the largest producers of GHG emissions. Thanks to the use of organic fertilizers and the inclusion of greater proportion of leguminous plants in crop rotations, the organic farming can contribute significantly to the emission load formation. Thanks to these measures, mainly N_2O emissions are reduced, while the N_2O is identified as a major greenhouse gas and its effect on climate is often referred to as 300 times greater as compared with the effect of CO_2. Another positive aspect of organic farming in terms of greenhouse gas emissions is the reduction of the number of animals per unit of area and limitation of point load caused by high concentrations of animals in one place which is typical for intensive industrial agriculture. Extensiveness of livestock production within organic farming system leads to reduction of methane production and in addition leads to further positive effects on soil and water quality, and in the broader context, also on biodiversity.

Measures	Impact
Fertilization	Using leguminous plants in crop rotations for the fixation of nitrogen and using organic fertilizers replace the use of synthetic fertilizers and the capacity of the soil for carbon sequestration is increased.
Protection against weeds	Thanks to the emphasis on the structure of crop rotations, mechanical, biological and other non-chemical methods of plant protection, the application of herbicides is eliminated.
Protection against pests	Thanks to the selection of resistant varieties, crop rotation edition, use of cover crops, intercrops and undersowing and support of predators and antagonists, the use of insecticides is reduced.
Protection against fungi and mildew	Due to the cultivation of resistant varieties, changes of crop rotation structure, emphasis on seed quality and the use of non-chemical methods of protection, the use of fungicides is reduced or even eliminated.
Closed farm cycle	Ensuring the maximum share of feed on the farm and the correct management of the herd minimize the need to purchase feed.
Continuous soil cover	Minimizing of periods without vegetation cover helps to increase the content of soil organic matter and its decomposition which reduces the need for fertilization.

Table 2. Tools for reducing the emission load resulting from the specifics of organic farming (adapted according to [34])

In terms of reducing greenhouse gas emissions, the another benefit of organic farming is the fact that organically cultivated arable land stores more carbon into humus. Thus the increase in atmospheric CO_2 is limited and this contributes to the climate stabilization. Binding of carbon dioxide is significantly higher in a longer crop rotation with perennial legume-grass mixture and with fertilization with manure. This is due to the increasing humus content in the soil, longer green land cover with catch crops and more powerful root system of main crops [45]. Rodale Institute's Farming Systems Trial states that the introduction of organic farming nationwide in the USA would manage to reduce CO_2 emissions by up to a quarter due to increased carbon sequestration in soils [32]. Emissions of carbon dioxide from organic farms is up to 50% lower per hectare. The balance of carbon dioxide is positively influenced by non-use of synthetic nitrogen fertilizers and pesticides and also by low doses of phosphorus and potassium, as well as low doses of grain fodder [41].

1.5. Biodiversity

The positive impact of organic farming on biodiversity is based on an effort to extend the range of cultivated crops and livestock and thereby to increase a genetic, species and ecosystem diversity. On this basis and on the basis of environmentally friendly agroecosystem management, the functional agrobiodiversity has increased. Growing biodiversity at all levels (predators, parasites, wild plans, pollinators, soil fauna and flora…) supports the ecosystem functions (population control, competition, allelopathy, organic matter decomposition, nutrient sorption and their cycle…). It contributes to the agroecosystem stability and sustainability. It improves resistance of production organisms against harmful agents gradation and contributes to their effective control, improves the nutrient utilization in agroecosystem and reduces eutrophication. It contributes to the erosion reduction and improves the moisture use, helps to increase diversity and abundance of wild flora and fauna in the landscape.

The high degree of diversity in the landscape, including agricultural land, can be caused by either a variety of abiotic environment (e.g. altitude, height relief zoning, seat rock and soil cover) or by disruption, disturbance caused by both natural interference processes and human activities. In the landscape structure, we can distinguish large areas whose internal environment a limited number of specialized species-"interior species" are bound to and smaller flats, transition zones and various broad interfaces that generate a colourful environment with many species corresponding to a diversity of ecosystems. On the edges of the fields of organic farmers, there are by 25% more birds surveyed than within conventional farming-in autumn and winter even 44% more [11]. They include both species characteristic of the individual habitats-species from forest, field, meadow and species from marginal environments-"edge species"-ecotone species that require more landscape elements for their existence. On the fields in organic farming, there are more accompanying plant species grown, in the ground layer 20-400% more species of wild plants [28]. Among other things, many endangered species of weeds as well [47].

Beneficial organisms prefer natural areas adjacent to the organic fields. Natural areas adjacent to the organically cultivated areas significantly support more beneficial organisms (e.g. ground beetles, spiders, wolf spiders-Lycosidae family and others for nature conservation significant fauna species) than the natural areas adjacent to the integrated areas or areas under the extensive farming [44]. In 41 of 45 studies, the number of earthworms, ground beetles, spiders (especially the Wolf spiders) and birds in the cultural landscape was significantly higher in organically cultivated areas than in the conventional ones. In four cases, there was no difference experienced [43].

The high degree of biodiversity reduces the population of rare species in the inner zones, increases the population of species in the border zones and of animals that require more landscape elements. From this, we can deduce that the right long-term agroecosystem function is directly proportional to the degree of biodiversity and the appropriate degree of stability of the area as a whole.

There are many factors driving agrobiodiversity such as crop range widening (broad crop rotation, specific or varietal mixtures, catch crops, cover crops, intercropping…), supply of organic matter into soil (manure, green manure, perennial crops), optimal fertilization, plant nutrition and protection, soil conservation technologies (direct drilling, mineralization of soil cultivation…) as well as the landscape feature creation (cops, alleys, strip planting, land division…). Areas within organic farming are more diverse (heterogeneous). Organic farmers farm on smaller fields with a greater proportion of green areas and a greater number of plant species. There are also more hedgerows in the organic farm.

In the organically cultivated farms, there are more than 85% greater number of plant species, a third more bats and there live about 17% more spiders and 5% more bird species [20, 23]. On the organic field, there are for example nine times more species of plants and accompanying weeds growing, there live 15% more ground beetles and 25% more earthworms than in the fields within integrated farming [35]. Greater variety of plants, hedges, grassy field margins, smaller areas of land, smaller corn ear density, area gardening, stubble and green soil cover in winter create favourable conditions for e.g. skylark. Already after one year of transitional period, the number of skylark nests has doubled. Nesting swallows and birds of prey also give priority to food from organic areas. In autumn and in winter, there were significantly more seeds and insects for songbirds and also more food for birds of prey found on organically cultivated fields [24].

2. Effect of farming system on greenhouse gas emissions

Among the positive externalities of organic farming belongs its environmental friendliness. This is also evident in the ability to produce less greenhouse gas emissions as compared to conventional intensive farming systems. In order to compare different farming systems more accurately, it is necessary to make a comprehensive analysis of materially energy flows within them and to quantify their impact on the environment. There can be used for example the LCA analysis (Life Cycle Assessment).

LCA analysis is a tool that enable to assess environmental impacts via the product life cycle. Within its framework, we can include also social and economic aspects but the main focus is on the environmental component. It is also an invaluable tool in the assessment of greenhouse gas emissions related to product formation [18].

The LCA study consists of four basic stages: Definition of objectives and the scope, Inventory, Impact assessment and Interpretation [27]. According to [29], in the first step when implementing the LCA methodology (goal setting), the reason for carrying out the method is specified. Life cycle inventory consists in simulating of a product system. Based on the knowledge of the life cycle of the product under consideration and on the basis of previously set system boundaries, there are first all involved processes and their inputs and outputs identified. Connecting processes with adequate energy and material flows into a functional complex, we obtain a product system diagram [26]. In essence, it is a qualitative and quantitative inventory of all inputs and outputs connecting the monitored system with the environment or the collection of the necessary primary data and an assessment of their quality, i.e. authenticity, reproducibility, transparency and confidentiality [29].

The aim of the life cycle impact assessment is to measurably compare the environmental impacts of product systems and to compare their severity with new quantifiable variables identified as impact category [27]. It is basically a qualitative and quantitative assessment of all negative effects that may be caused in the environment by influences collectively specified in the inventory matrix [46]. In the interpretation phase, outputs of LCA analysis are described. During the inventory phase and the phase of impact assessment, there were made some certain estimates, assumptions and judgements how to continue with the study. There have been some simplifications and approximations adopted. All of these elements must be included in the interpretation phase and must always be set next to the result presentation [27].

For the correct execution of the LCA analysis, it is important to determine the boundaries within which the particular life cycle processes will be monitored. Determination of system boundaries is always a very important step, especially in the area of food production and agriculture, where the clearly identifiable technological processes and systems meet the natural processes and procedures influenced by a number of factors [2, 3].

Each product system consists of a variable number of processes involved in the product life cycle. However, the product under consideration is often related to other processes that may no longer be important for the LCA study. The system boundary serves to the separation of essential and non-essential processes of the product life cycle. Since the choice of system boundaries significantly affects LCA study outcomes and in addition, its intensity and complexity, system boundaries should always be well considered and clearly defined. The choice of system boundaries is carried out with regard to the studied processes, studied environmental impacts and selected complexity of the study.

Not-including any life cycle stages, processes or data must be logically reasoned and clearly explained [26].

When comparing conventional and organic farming systems, we can also omit the calculation of load from buildings and infrastructure because there are only small differences between farming systems while slightly more noticeable difference is apparent within animal production [40]. The reduction of processes under consideration can be made also on the basis of their presumed significance. The assessment can skip those processes, whose overall impact consists totally of only a negligible share, for example of only several percent. These processes can be identified for example based on a comparison of similar studies or expert estimation but in this process it is always necessary to take into account the specifics of the particular situation. If the process significance is set incorrectly, this could have a rough influence on the results of the whole analysis.

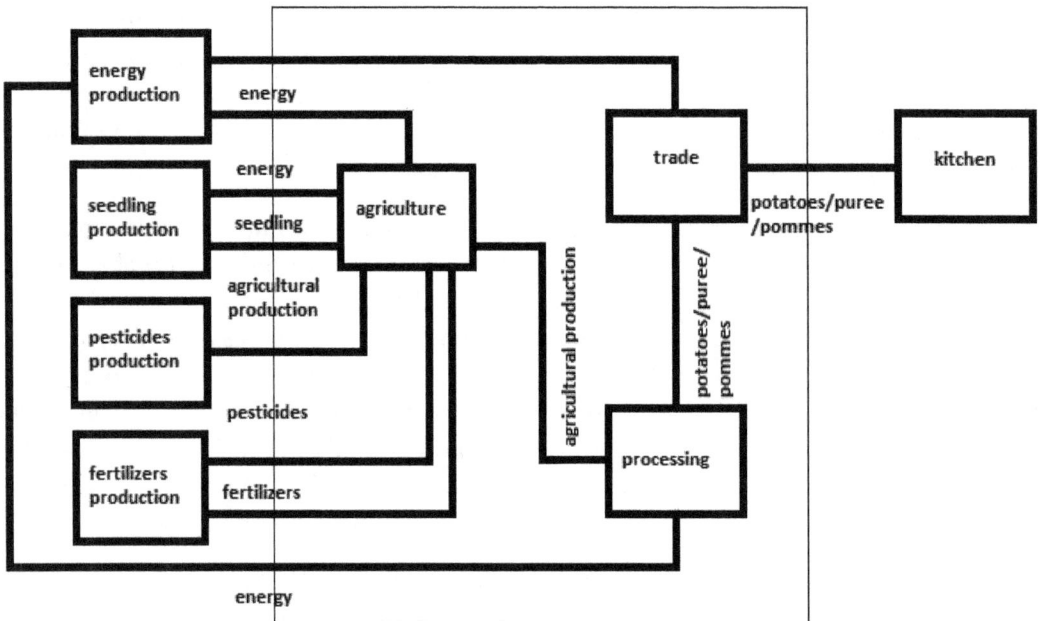

Scheme 3. Example of setting the boundaries of the system within LCA analysis [37]

To compare products (systems), it is necessary to define also the functional unit. The functional unit is described as a quantified performance of a product system which serves as a reference unit in a study of life cycle assessment [13]. It is an essential element which all study results are related to. It must be chosen so as to be easily expressible and measurable. The functional unit is the starting point for searching for alternative ways how to fulfil the function with a lower negative impact on the environment [55]. [1] states that the determination of functional units is as a crucial step especially when comparing systems with different levels of production per hectare such as conventional and organic farming system. [15] recommends to involve both functional units into calculations and perform the calculations for both the unit area and the unit of production. The impact of organic system on the mitigation is usually measured per unit area in order to enhance the objectivity. However, it is important to convert it also to the production unit. Greenhouse gas emissions are lower in organic systems per unit area but also per unit of production. However, environmental savings per unit area are due to lower yields within organic farming roughly double as compared with the calculation per unit of production [40]. [25] also states that due to lower yields within organic farming in the calculation of GHG emissions per unit of production, the environmental load increases in relation to conventional farming, so the resulting difference is less than when converted to the unit area. This is consistent with findings of [36] who states that due to lower yields in organic farming, particularly in less developed countries, the environmental effect consisting in the reduction of greenhouse gas emissions is lower when converted to unit of production instead of unit area and in extreme cases, it can even be negative. However, for both methods of calculation for most crops, the greenhouse gas emissions production within organic farming remains lower [38].

Greenhouse gas emissions are expressed in relation to their effect on climate changes by an equivalent of CO_2e ($CO_2e=1x\ CO_2+23x\ CH_4+298x\ N_2O$). Within various subphases of agricultural production cycle, the emission load of conventional and organic farming differs. For example in the agrotechnical phase with most cultivated crops, the higher emission load per one kilogram of production occurs within the organic farming system. However, it is usually more than satisfactorily compensated by the absence of synthetic fertilizers (especially nitrogen) which are, in terms of emission production, the most loading element. The way of fertilization is directly related to field emissions which are lower in organic farming and there are some savings also due to the absence of pesticide use. To sum up, organic farming produces lower GHG emissions within cultivation of most crops and in some cases, there are very fundamental differences (e.g. Conventional production of rye under condition of central Europe produces almost twice the emissions CO_2e in relation to organic farming).

This can be documented by the study within which the creation of greenhouse gases within the cultivation of crops in conventional and organic farming system under conditions of central Europe was compared. Within the study, the total GHG emissions expressed as CO_2e were observed. This sum was divided into subgroups-agricultural engineering, fertilizers, pesticides, seeds and field emissions. The conventional farming system differs from the organic one in the total CO_2e emissions production as well as in the production within subgroups. Overall results are shown in Figure 1 that summarizes the production of greenhouse gases converted to one kilogram of production and compares the conventional system with the organic farming system.

GHG emissions within cultivation of particular crops vary depending on many factors, while the most CO_2e is released within fertilization and field emissions and also a share of agricultural operation in not negligible. With all surveyed crops except onion, where 0.083 CO_2e/kg of onion in conventional and 0.100 kg CO_2e/kg of onion in organic farming is produced, higher CO_2e emissions were found within the conventional farming system. Within cultivation of wheat, 0.460 kg CO_2e/kg of grains within conventional and 0.423 kg CO_2e/kg of grains within organic farming is released. With rye, it is 0.537 kg CO_2e/kg of grains within conventional and 0.298 kg CO_2e/kg of grains within organic farming, with potatoes 0.145 kg CO_2e/kg of potatoes within conventional and 0.125 kg CO_2e/kg of potatoes within organic farming, with carrot 0.099 kg CO_2e/kg of carrot within conventional and 0.041 kg CO_2e/kg of carrot within organic farming, with tomatoes 0.087 kg CO_2e/kg of tomatoes within conventional and 0.067 kg CO_2e/kg of tomatoes within organic farming and with cabbage 0.078 kg CO_2e/kg of cabbage within conventional and 0.033 kg CO_2e/kg of cabbage within organic farming. It is obvious, that the organic farming system is, in terms of emission, less demanding and therefore more environmentally friendly than conventional farming, where emissions production is increased especially by the use of synthetic fertilizers.

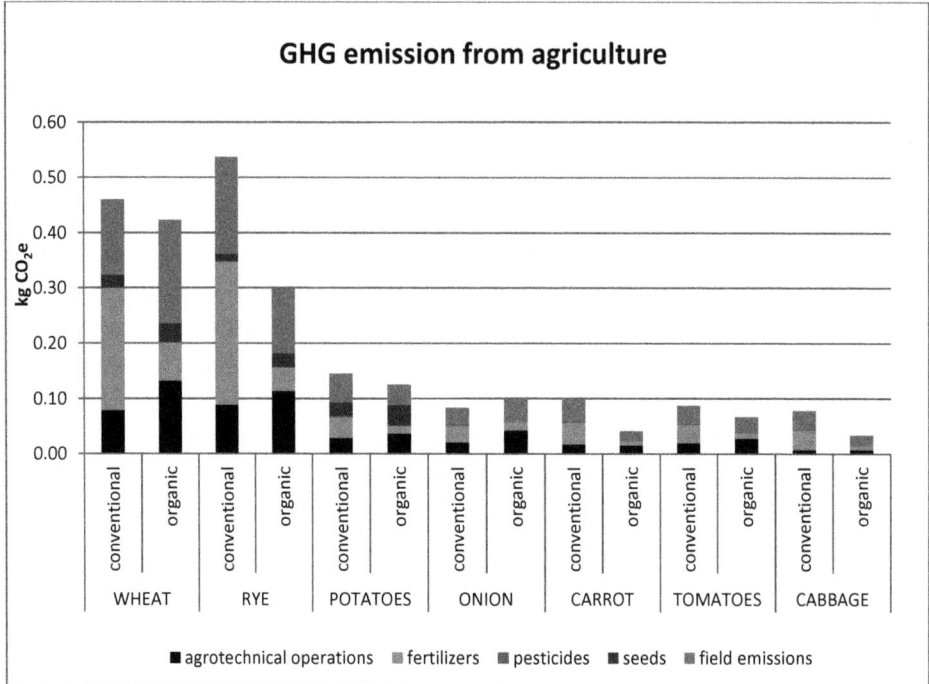

Figure 1. Total production of GHG emissions within growing of selected crops in conventional and organic farming system (in kg CO_2e per 1 kg of production) [38].

Differences between particular subgroups (Farming, Fertilizers, Pesticides, Seeds and Field emission) can be documented as exemplified by the comparison of the cultivation of wheat and rye in conventional and organic farming system. As stated by [31], the system sustainability can be evaluated on the basis of inputs and outputs and their conversion to CO_2e. Due to higher demands for agro-technical procedures and lower yield per hectare, GHG emissions generated in the organic farming system are with rye and wheat higher as compared with the conventional farming system. With wheat, as it is evident from Figure 2, the values within organic farming (0.132 CO_2e/kg of grains) are 69.2% higher than in the conventional system (0.078 CO_2e/kg of grains), with rye then higher by 28.4% (conventional agriculture 0.088 CO_2e/ kg of grains, organic farming 0,113 CO_2e/kg of grains). The possibility of reducing GHG emissions by changes in agricultural technology is highlighted also by [16] who identifies the main potential for reduction within tillage. Zero-tillage systems as a tool for reduction of the emission load is also mentioned by [31], who states that the change to zero-tillage systems can lead to reduction of emissions of 30-35 kg C/ha per a vegetation period. Also [4] stated that the technique of reduced (minimum) tillage which is in organic systems used more and more frequently and with greater success supports carbon sequestration significantly.However, unlike conventional zero-tillage systems, the organic systems with limited (minimum) tillage do not require higher inputs of herbicides and synthetic nitrogen [42]. Parallel to the increase of yields, the minimization measures in agricultural technology can contribute to the reduction of emissions and thereby further to increase the environmental emission savings ensured by the organic farming system.

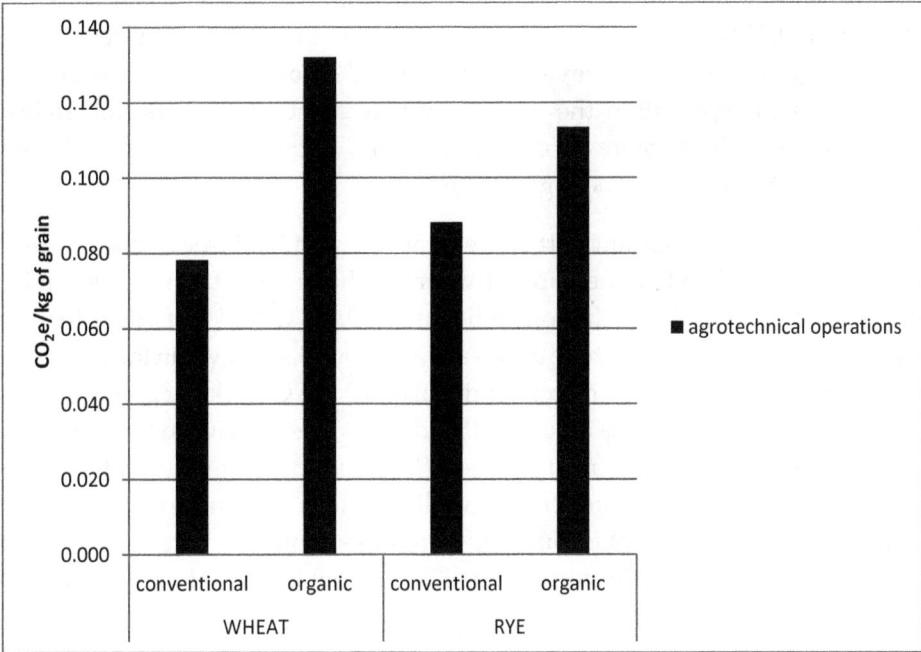

Figure 2. Production of CO_2e/kg of grain from agrotechnical operations during growing of wheat and rye in organic and conventional system of farming

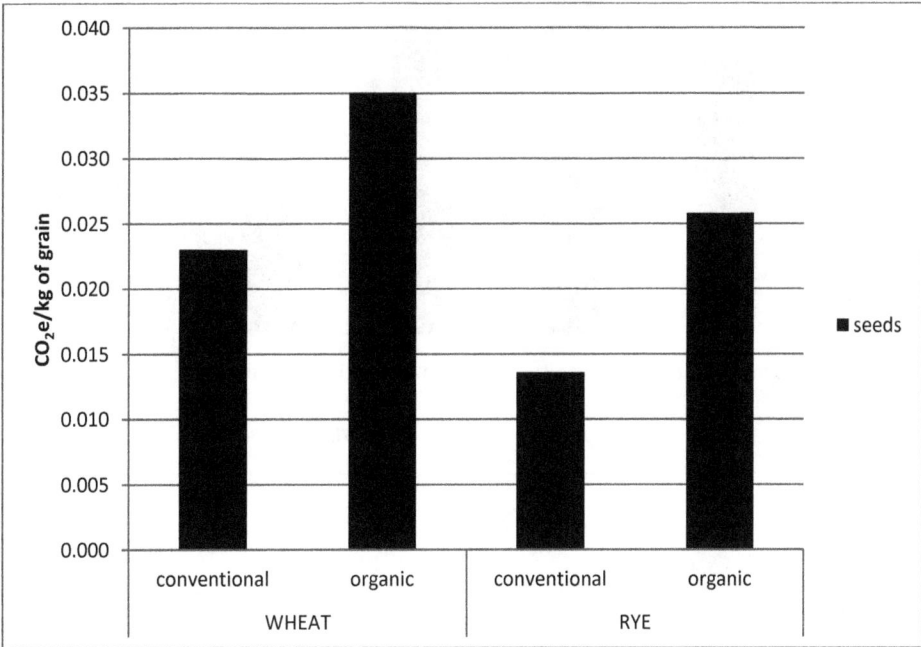

Figure 3. Production of CO_2e/kg of grain from seeds during growing of wheat and rye in organic and conventional system of farming

Higher GHG emissions are within the organic system also with seed (see Figure 3), where the calculated value is 0.023 CO_2e/kg of grains for wheat within the ecological system, respectively 0.035 CO_2e/kg of grains within the conventional one (52.2% more in organic farming) and 0.014 CO_2e/kg of grains for rye within the organic system and 0.026 CO_2e/kg of grains in the conventional system (85.7% more in organic farming). However, of the total amount of emissions, the seed has relatively negligible share.

Field emissions make an important part of the total emission load (see Figure 4), there is not evident an explicit trend. While they are for wheat in the organic farming system by 36.5% higher (0.137 CO_2e/kg of grains in the conventional and 0.187 CO_2e/kg of grains in the organic farming system) however, for rye, there are these values by 33.7% lower in the organic farming system (0.175 CO_2e/kg of grains in the conventional and 0.116 CO_2e/kg of grains in the organic farming system). Differences in the amount of field emissions are caused by a combination of several factors, the main role is played by the difference in the intensity of fertilization (for wheat in the organic farming system, there is applied 20 tons of manure while for rye only 12 tonnes) and by the various ratio of income between conventional and organic farming system (organic wheat yield is 58% of the conventional wheat yield, yield of organic rye makes up 72.5% of the conventional rye yield).

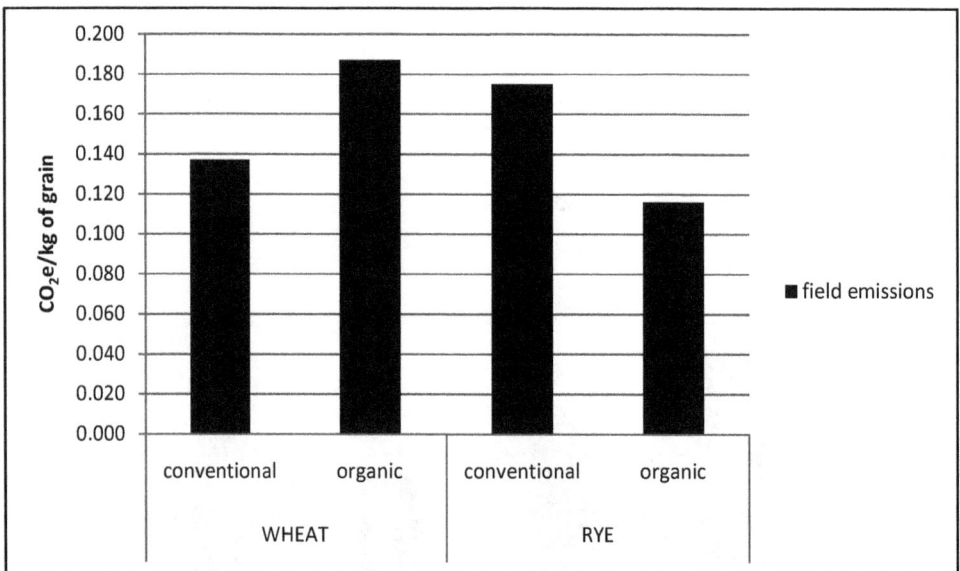

Figure 4. Production of CO_2e/kg of grain from field emissions during growing of wheat and rye in organic and conventional system of farming

The main differences in emission production in conventional and organic farming system arising from the use of agrochemicals in the conventional farming system. While there is a relatively small share of total emissions caused by the use of pesticides (0.22% of total emissions for wheat and 0.19% for rye) and their environmental impact lies mainly in their toxicity, a very high emission load is generated by using synthetic fertilizers (conventional farming system) instead of organic fertilizers (organic farming system). This is consistent with findings of [5, 19] who state that synthetic fertilizers are the main source of greenhouse gas emissions. As it is evident from Figure 5, the conventional farming produces in this stage of the process significantly more GHG emissions. With wheat, these values (0.221 CO_2e/kg of grains in the conventional farming system, 0.069 in the organic farming system) are in organic agriculture by 68.8% lower, for rye then 83.4% lower (0.259 CO_2e/kg of grains in the conventional farming system, 0.043 in the organic farming system).

GHG emissions from fertilization make up a value around 48% (48.04% wheat, 48.27% rye), this is the largest share of total emissions in conventional agriculture, which is consistent with findings of [21] who states this proportion of 40-50% for rape and approaches the data by [6] who states the range of 35-40%. While in organic farming, emissions from fertilization make up in the total amount of GHG emissions only 16.31% with wheat and 14.41% with rye. [17] states that nitrogen management in agriculture loses its effectiveness in terms of the proportion of utilization of applied nitrogen. The total number of inputs is increasing but plants consume actually still smaller share of the applied nitrogen. A large part of the increased amount of fertilizer is not processed by the plant but is released into the water or into the air. To reduce greenhouse gas emissions, it is necessary next to a reduction of synthetic fertilizers also a proper management of their application or the application of fertilizers in general. Both within conventional and organic farming system, there should be able to reduce the environmental load arising from fertilization with nitrogen while maintaining current yield levels.

In terms of emission production, the extensification of conventional farming or its conversion to the organic farming system can be the step leading to a reduction in the overall proportion of anthropogenic GHG emissions. On the contrary, within organic farming, the increased yields are seek. They emphasizes the environmental friendliness also after conversion to the unit of production.

The increase of income, while maintaining the current structure of inputs of organic farming, as a way of deepening its environmental benefit, is referred also by [9]. This is consistent with findings by [12] who states that organic farming systems are significantly more environmentally friendly when they reach a relatively higher yields. [7] states that this can be achieved by e.g. more efficient application of fertilizers and crop rotation balance. Also, the proper selection of varieties could significantly contribute to better yields and their stability within organic farming and in the farming system with low inputs [8]. Productivity in sustainable farming can be increased through many indirect measures based on improving of soil fertility and stimulation of the functions of plants and microorganisms in natural soil processes. The most important role in the soil is yet played by carbon. It is important for soil moisture and at the same time thanks to increasing the carbon content or the soil organic matter in the soil, the production of greenhouse gases released into the atmosphere can be reduced. Strengthening these soil processes in order to increase the productivity is typical for organic farming [40].

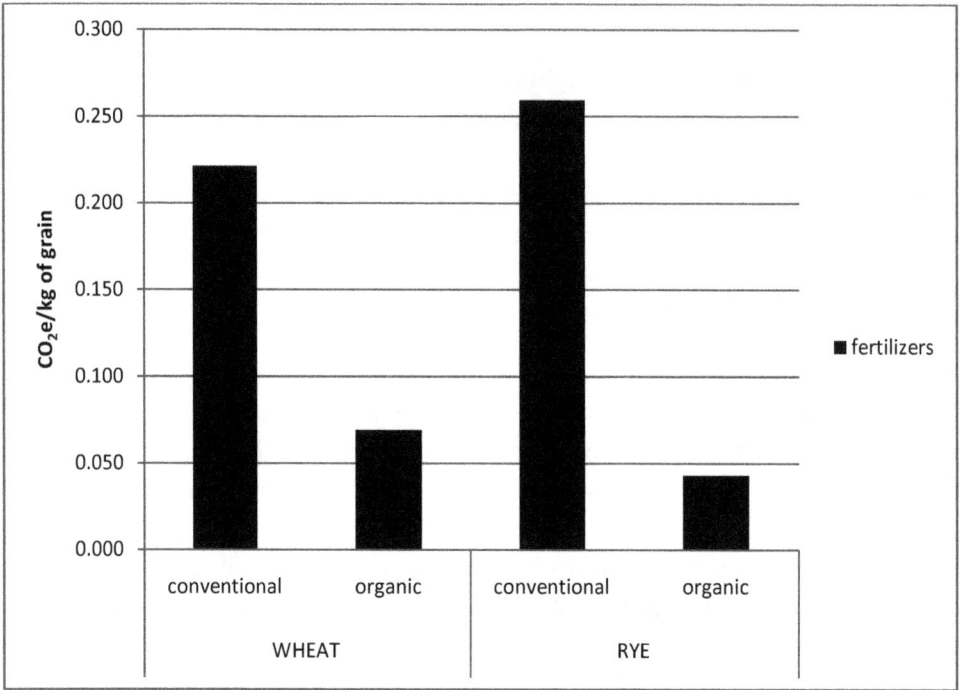

Figure 5. Production of CO_2e/kg of grain from fertilizers during growing of wheat and rye in organic and conventional system of farming

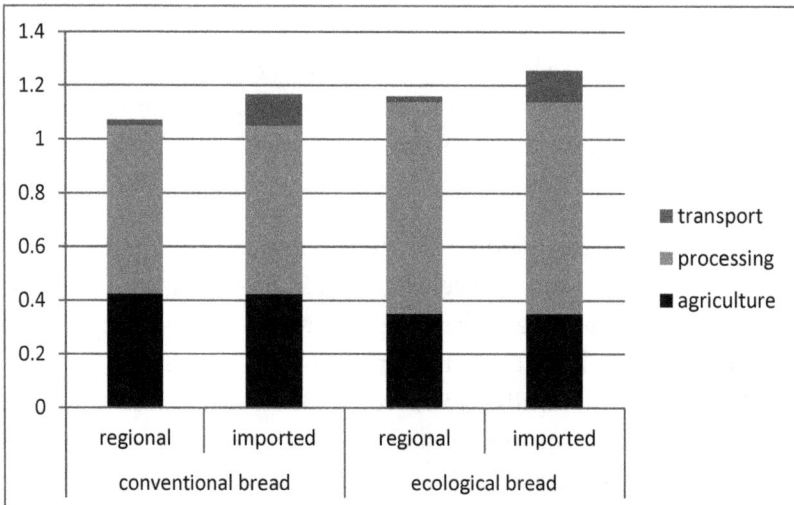

Figure 6. Emission load of bread (in kg CO_2e per 1 kg of bread) [39]

In addition to the agricultural production, also other processes such as processing or transport may be incorporated within the framework of the LCA analysis. While within the livestock production, the post-farm phase is less important in terms of emissions, for crop production, its importance is increasing [52]. As it is evident from figure 6, these processes may in some cases reverse the environmental potential created by the organic farming system within the farm phase. In the case of the production of bread in bio-quality under conditions of the Czech Republic, this leads to a situation where due to different technologies and especially production capacity in organic farming, the final product produces in terms of CO_2e emissions a higher load than conventional products.

Emission load from the transport is in the case of bio-products also often increased due to a lower density of organic producers and processors, thereby the average transport distance between the production chain elements is longer. However, despite the above factors, with a number of products, the greenhouse gas emissions per unit of production remain lower even for the final product. E.g. for the cultivation of potatoes and mashed potatoes production in organic farming, the emission load remains slightly lower even after including post-farm part of the cycle.

3. Conclusion

Predominantly, the farming systems have still been considered as production systems. The typical conventional agroecosystems are characterized as very open, as systems based on technological processes substituting the ecological ones, systems with high labour productivity but with lower biodiversity, flexibility, stability and sustainability on the other hand. On the contrary, the organic farming systems are based on sustainable development principles and the holistic world approach. It is a production system focused on preservation and improvement of natural resources and the environment at the same time. As for the system concept, there is an effort to balance the economic, environmental and social aspects and relations on global and local level. Agricultural activity itself is considered a process of reasonable ecosystem exploitation with respect to its stability and sustainability. Just this divergence between the both approaches, the intensive industrial agriculture on one hand and the sustainable organic farming systems on the other, makes them very different from the environmental point of view.

Intensified conventional farming leads to soil quality decrease. Intensive soil cultivation affects the soil structure negatively and raises the risk of soil erosion. Because of the substitution of mineral fertilizers for organic manures, the content and quality of organic matter decreases and the microbial activity is disturbed. Within the effort to increase the labour productivity, there has been often used larger and heavier agricultural machinery, which leads to intensive soil compaction resulting in unbalanced soil air and water regime, limited root system development, soil biological activity and water absorption reduction and erosion risk increase. The external technical and material inputs markedly increase the energy demands and thus consumption of non-renewable resources together with higher atmospheric pollution. As the specialization develops, number of used species decreases, the preventive function of cropping patterns (pests and diseases reduction) degrades. Crop breeding for high yields brings higher demands for fertilizing and higher plant sensitivity to adverse environment conditions. There is also an increased need of synthetic nitrogen fertilizers. Residual content of nitrogen contributes to underground and surface waters pollution and nitrogen evaporation into the atmosphere. The accumulation of such active substances in the soil results in destruction of useful microorganisms, antagonists and other soil organisms, also leads to development of resistance to pesticides in harmful organisms, decrease of plant and animal species number, pollution of underground and surface waters and the atmosphere with a negative impact to the whole ecosystem. Similarly, the intensifying concentration and specialization of livestock production results in great local amounts of organic wastes, possibilities of their utilisation are not sufficient and the risk of soil and water pollution rises.

Within the organic farming systems the soil quality remains the main interest. Soil erosion control measures include cover crops, mulching, limited soil cultivation, windbreaks planting, use of lighter and smaller machinery, keep an optimal soil structure and looseness. These preservative soil cultivation principles are combined with lower need of pesticides. Thus there is ensured a sufficient nutrient cycle and organic matter content in the soil leading to an optimal soil biological activity and fertility. Generally, the principles of organic farming ensure protection of water sources and soil moisture, prevent the underground and surface waters from pollutants and sediments as well. Water preservation is the priority, there are used terraces, environmental corridors and border zones and other measures. There are also considerable differences in biodiversity, which, within the conventional farming, markedly suffers not only due to crop range reduction leading to monocultures in fact, but also thanks to reduction of associated fauna and flora thought as harmful and thus systematically eliminated or suffering from pesticides or other biocide substances at the same time. By contrast, organic farming purposefully supports biodiversity, takes advantage of more adaptable animal and crop species and varieties to the habitat conditions, uses varied crop rotations, species and variety mixtures, applies technical and organizational measures friendly to the organisms and the environment. Organic farming systems are also more environmentally friendly with regard to the greenhouse gases emissions production. Above all, this emission reduction is achieved thanks to the limited use of synthetic fertilizers and pesticides and lower livestock production intensity.

Although the organic farming systems show much friendlier environmental impact when compared to the conventional ones, there has still been room for further improvements. When compared to the conventional farming systems, one of the most important weakness of contemporary organic farming is a low production capability. In countries with established organic farming yields on arable land reach only 45-100 % in comparison to conventional farming. This difference shows a specific reserve and rising potential of the production capability through appropriate intensification related to natural fertility of a habitat. Among the environmentally acceptable (rational, biological, technological…) means and methods of the ecofunctional intensification within organic farming systems belong e.g.:

• Better management of the soil organic matter

• Use of perennial leguminous plants

• Support of the soil-plant interaction

- Edaphon environment optimisation

- Intensified recycling and better utilisation of macro and micronutrients, mycorrhiza and nitrogen fixation

- Integration of food production and energy (intercrop biogas production, green manure, stable dung and dung water fermentation prior to their recirculation)

- Landscape management leading to the biodiversity improvement

- Improvement of technology and products used for weeds, pests and diseases elimination (e. g. biological protection, herbal-based pesticides, allelopathy and physical barriers)

- Animal and plant breeding based on the accent on preservation of indigenous markers for resistance and multifunctionality and suitability for organic farming

- Implementation of the precise farming principles into the organic farming systems (automation and robotization, use of sensors in crop and livestock production, GPS and IT)

- Development of new techniques and skills within crop and livestock production that comply with ecological principles and standards (e. g. intercropping, polycultures...)

Sustainable systems should be more focused on preventive measures (crop rotation, precise variety selection), biological regulation methods and balance of all factors of crop production. Similarly, for the conventional farming systems, there can be implemented measures leading to decrease in their environmental impact. Among the main proecological measures belong:

- Land use and a company structure optimisation correspondent to a locality

- Provide maximal species diversity

- Follow crop rotation principles within cropping patterns

- Optimization of the share of leguminous plant related to the soil nutrient content balance and feedstuff need

- Take into account the weeds and pests damage thresholds for pesticides reduction

- Use of mechanical, biological and organizational methods within the plant and animal protection

- Preservation of good soil structure by the means of timely applied field operations

- Reduction and modification of technological operations

- Regular manuring and sweetening

- Reduction of the area under wide-row crops

- Preserve the green soil cover as long as possible

- Use of intercrops, undersowings and green belts

- Preservation of meadows and pastures in flooded areas and slopping grounds

- Preservation of field boundaries reducing soil erosion

- Landscape element planting

The negative impact of intensive conventional farming on the environment and production quality has been gradually reduced by implementation of directives applied by EU members. Public requirements on farming related to all environmental components is predominantly covered by the Crosscompliance system, which emphasises the importance of a responsible control of quality and food safety and animal welfare. The financial support of farming depends on fulfilling conditions stated in GAEC (Good Agricultural and Environmental Conditions), on SMR (Statutory Management Requirements) and on minimal requirements on fertilizer and pesticides use in compliance with the Agroenvironmetal measures (AEO). The mentioned directions generally accepts above stated requirements for improvement of the environmental aspect of the conventional farming.

Acknowledgements

Our work was supported by the Ministry of Agriculture of the Czech Republic – NAZV, Grant No. QJ1310072 and University of South Bohemia in České Budějovice - GAJU, Grant No. 063/2013/Z.

Author details

Jan Moudrý jr. and Jan Moudrý*

*Address all correspondence to: JMoudry@zf.jcu.cz

Faculty of Agriculture, University of South Bohemia in České Budějovice, České Budějovice, Czech Republic

References

[1] Basset-Mens C., Van Der Werf HMG. Scenario-based environmental assessment of farming systems: the case of pig production in France.Agriculture, Ecosystems & Environment 2005; 105127-144.

[2] Berlin D., Uhlin HE. Opportunity cost principles for life cycle assessment: Toward strategic decision making in agriculture.Progress in Industrial Ecology, An International Journal 2004; 1 187-202.

[3] Berlin J. Environmental life cycle assessment (LCA) of Swedish semi-hard cheese.International Dairy Journal 2002; 12939-953. Berner A., Hildermann I., Fließbach A.,

Pfiff ner L., Niggli U., Mäder P. Crop yield and soil fertility response to reduced tillage under organic management. Soil & Tillage Research 2008; 101 89–96.

[4] Berner A., Hildermann I., Fließbach A., Pfiff ner L., Niggli U., Mäder P. Crop yield and soil fertility response to reduced tillage under organic management. Soil & Tillage Research 2008; 101 89–96.

[5] Biswas WK., Barton L., Carter D. Global warming potential of wheat production in South Western Australia: a life cycle assessment. Water Environment Journal 2008; 22206-216.

[6] Braschkat J., Braschkat A., Quirin M., Reinhardt GA. Life cycle assessment of bread production-a comparison of eight different scenarios. DIAS report,Animal Husbandry2004; 61 9-16.

[7] Brentrup F., Küsters J., Kuhlmann H., Lammel J. Environmental impact assessment of agricultural production systems using the life cycle assessment methodology-I. Theoretical concept of a LCA method tailored to crop production.European Journal of Agronomy 2004; 20 247-264.

[8] Burger H., Schloen M., Schmidt W., Geiger HH. Quantitative genetic studies on breeding maize for adaptation to organic farming. Euphytica 2008; 163 501–510.

[9] Cederberg C., Mattsson B.Life cycle assessment of milk production-a comparison of conventional and organic farming.Journal of Cleaner Production 2000; 849-60.

[10] Cerri CC., Maia SMF., Galdos MV., Cerri CEP., Feigl BJ., Bernoux M. Brazilian greenhouse gas emissions: the importance of agriculture and livestock. Scientia agricola2009; 6 831-843.

[11] Chamberlain DE., Wilson JD., Fuller RJ. A comparison of bird populations on organic and conventional farm systems in southern Britain. Biological Conservation 1999; 88 307–320.

[12] Charles R., Jolliet O., Gaillard G., Pellet D. Environmental analysis of intensity level in wheat crop production using life cycle assessment.Agriculture, Ecosystems & Environment 2006; 113 216-225.

[13] ČNI 2006. ČSN EN ISO 14 040: Environmentální management – Posuzování životního cyklu – Zásady a osnova. Praha: Český normalizační institut; 2006.

[14] Cole CV., Duxbury J., Freney J., Heinemeyer O., Minami K., Mosier A. et al. Global estimates of potential mitigation of greenhouse gas emissions by agriculture. Nutrient Cycling in Agroecosystems 1997; 49 221-228.

[15] De Backer E., Aertsens J., Vergucht S., Steurbaut W. Assessing the ecological soundness of organic and conventional agriculture by means of life cycle assessment (LCA): A case study of leek production. British Food Journal 2009; 111 1028-1061.

[16] Dyer JA., Desjardins RL. The impact of farm machinery management on the greenhouse gas emissions from Canadian agriculture. Journal of Sustainable Agriculture 2003; 22 59-74.

[17] Erisman JW., Sutton MA., Galloway J., Klimont Z., Winiwarter W. How a century of ammonia synthesis changed the world. Nature Geoscience 2008; 1 636–639.

[18] Finnveden G., Hauschild MZ., Ekvall T., Guninée J., Heijungs R., Hellweg S. et al.Recent developments in life cycle assessment. Journal of Environmental Management 2009; 91 1-21.

[19] Fott P., Pretel J., Vácha D., Neužil V., Bláha J. Národní zpráva České republiky o inventarizaci emisí skleníkových plynů. Praha: ČHMÚ; 2003.

[20] Fuller RJ., Norton LR., Feber RE., Johnson PJ., Chamberlain DE., Joys AC. etal. Benefits of organic farming to biodiversity vary among taxa. Biology Letters 2005; 5187-202.

[21] Gasol CM., Gabarell X., Anton A., Rigola M., Carrasco J., Ciria P. et al. Life Cycle Assessment of a Brassica carinata CroppingSystem in Southern Europe. Biomass Bioenergy 2007; 31 543–555.

[22] Haas G., Berg M., Köpke U. Nitrate leaching: comparing conventional, integrated and organic agricultural production systems. In: Steenvoorden J.,Claessen F., Willems J. (eds.) Agricultural Effects on Ground and Surface Waters: Research at the Edge of Science and Society. Oxfordshire: IAHS; 2002. p131-136.

[23] Hole DG., Perkins AJ., Wilson JD., Alexander IH., Grice PV., Evans AD. Does organic farming benefit biodiversity? Biological Conservation 2005; 122 113-130.

[24] Hötker H., Rahmann G., Jeromin K. Positive Auswirkungen des Okolandbaus auf Vogel der Agrarlandschaft – Untersuchungen in Schleswig-Holstein auf schweren Ackerboden. Sonderhefte der Landbauforschung Völkenrode 2004; 272 43-59.

[25] Knudsen MT. Environmental assesment of imported organic products-Focusing on orange juice from Brazil and soyabeans from China.Aarhus: Aarhus University; 2010.

[26] Kočí V. Na LCA založené srovnání environmentálních dopadů obnovitelných zdrojů energie-Odhad LCA profilů výroby elektrické energie z obnovitelných zdrojů energie v ČR pro projekt OZE-RESTEP. Praha: VŠCHT;2012.

[27] Kočí V. Příručka základních informací o posuzování životního cyklu. Praha: VŠCHT; 2010.

[28] Köpke U. Umweltleistungen des Okologischen Landbaus.Okologie & Landbau 2002; 2 6-18.

[29] Kotovicová J., Holešovská Z., Labodová A., Remtová K.Čistší produkce. Brno: Mendelova zemědělská a lesnická univerzita; 2003.

[30] Kvítek T, Tippl M. Ochrana povrchových vod před dusičnany z vodní eroze a hlavní zásady protierozní ochrany v krajině. Praha: Ústav zemědělských a potravinářských informací; 2003.

[31] Lal R. Carbon emission from farm operations. Environment International 2004; 30 981-990.

[32] LaSalle T. Regenerative Organic Farming. Pensylvania: Rodale Institute; 2008.

[33] Liess M., Schulz R., Berenzen N., Nanko-Drees J., Wogram J. Pflanzenschutzmittel-Belastung und Lebensgemeinschaften in Fliessgewassern mit landwirtschaftlich genutztem Umland. Berlin: Technische Univerität Braunschweig; 2001.

[34] Little T. Monitoring and management of energy and emissions in agriculture. Shropshire: Institute of Organic Training and Advice; 2007.

[35] Mader P., Fliessbach A., Dubois D., Gunst L., Fried PM., Niggli U. Soil Fertility and Biodiversity in Organic Farming. Science 2002;296 1694-1697.

[36] Mondelaers K., Aertsens J., Van Huylenbroeck G. A meta-analysis of the differences in environmental impacts between organic and conventional farming. British Food Journal 2009; 111 1098-1119.

[37] Moudrý J. jr., Jelínková Z., Jarešová M., Plch R., Moudrý J., Konvalina P. Assessing greenhouse gas emissions from potato production and processing in the Czech Republic. Outlook on Agriculture 2013; 42 179–183.

[38] Moudrý J. jr., Jelínková Z., Moudrý J., Bernas J., Kopecký M., Konvalina P. Influence of farming systems on production of greenhouse gas emissions within cultivation of selected crops. Journal of Food, Agriculture & Environment 2013, 3&4 (11) 1015-1018.

[39] Moudrý J. jr., Jelínková Z., Plch R., Moudrý J., Konvalina P., Hyšpler R. The emissions of greenhouse gasses produced during growing and processing of wheat products in the Czech Republic. Journal of Food, Agriculture & Environment 2013; 1 (11) 1133-1136.

[40] Nemecek T., Erzinger S. Modelling representative life cycle inventories for Swiss arable crops. International Journal of Life Cycle Assessment 2005; 10 68-76.

[41] Nemecek T., Kufrin P., Menzi M., Hebeisen T., Charles R. Okobilanzen verschiedener Anbauvarianten wichtiger Ackerkulturen. VDLUFA-Schriftenreihe 2002; 58 564-573.

[42] Niggli U., Fliessbach A., Hepperly P., Scialabba, N. Zemědělství s nízkými emisemi skleníkových plynů. Olomouc: Bioinstitut; 2011

[43] Pfiffner L., Häring A.,Dabbert S.,Stolze M.,Piorr, A.Contributions of Organic Farming to a sustainable environment. In: Bjerregaard R. (ed.) European Conference: Organic Food and Farming-Towards Partnership and Action in Europe, 10-11 May 2001, Copenhagen, Denmark. Denmark: Norhaven A/S, 2001.

[44] Pfiffner L., Luka H. Effects of low-input farming systems on carabids and epigeal spiders in cereal crops – a paired farm approach in NW-Switzerland. Basic and Applied Ecology 2003; 4 117-127.

[45] Pimentel D., Hepperly P., Hanson J., Douds D., Seidel R. Environmental, energetic, and economic comparisons of organic and conventional farming systems. Bioscience 2005; 55 573-582.

[46] Remtová K. Posuzování životního cyklu – METODA LCA. Praha:MŽP; 2003.

[47] Rydberg NT., Milberg P. A Survey of Weeds in Organic Farming. Biological Agriculture and Horticulure 2000; 18175-185.

[48] Šarapatka B., Urban J., Čížková S., Dukát V., Hejduk S., Hrabalová A. et al. Ekologické zemědělství v praxi. Šumperk: PRO-BIO; 2006.

[49] Schnug E., Haneklaus S. Landwirtschaftliche Produktionstechnik und Infiltration von Boden: Beitrag des okologischen Landbaus zum vorbeugenden Hochwasserschutz. Landbauforschung Volkenrode 2002; 52 (4) 197-203.

[50] Siegrist S., Schaub D., Pfiffner L., Mader P. Does organic agriculture reduce soil erodibility? The results of a long-term field study on loess in Switzerland. Agriculture, Ecosystems and Environment 1998; 69 253-265.

[51] Šimek M. Základy nauky o půdě 3. Biologické procesy a cykly prvků. České Budějovice: Biologická fakulta JU; 2003.

[52] Sonesson U., Davis J., Ziegler F. Food Production and Emissions of Greenhouse Gases. Göteborg: Swedish Institute for Food and Biotechnology; 2009.

[53] Stolze M., Piorr A., Häring A., Dabbert, S. The Environmental Impacts of Organic Farming in Europe – Organic Farming in Europe: Economics and Policy, vol. 6. Stuttgart: University of Hohenheim; 2000.

[54] Váchal J., Moudrý J. Projektování trvale udržitelných systémů hospodaření. České Budějovice: ZF JU; 2002.

[55] Weinzettel J. Posuzování životního cyklu (LCA) a analýza vstupů a výstupů (IOA): vzájemné propojení při získávání nedostupných dat. Praha: ČVUT; 2008.

2

Current Status of Advisory and Extension Services for Organic Agriculture in Europe and Turkey

Orhan Özçatalbaş

1. Introduction

Europe continent is covering more than 50 countries and the third most populous and the world's second-smallest continent by surface area of which 28 belong to the European Union (EU). The European Union reached its current size of 28 member countries with the accession of Croatia on July 1, 2013. The EU was not always as big as it is today. When European countries started to cooperate economically in 1951, only Belgium, Germany, France, Italy, Luxembourg and the Netherlands participated. Since that day, more and more countries decided to join the Union. As it known the European Union was built after the Second World War by the six countries who signed the Treaty of Rome in 1957. Since its creation European agricultural policy discussions have been dominated by the Common Agricultural Policy (CAP) has always been adapted to respond to the challenges of its time. From its start in 1962, the CAP begins to restore Europe's capacity to feed itself. In the 1980s production control measures begin. The CAP refocuses on quality, safety and affordability of food and on becoming greener, fairer and more efficient. The EU's role as the world's biggest trader in farm goods gives it additional responsibilities. A big CAP reform package in 1992 requires farmers to assume responsibility for environment protection and sustainable agriculture [13].

Significant reforms have been also made in recent years, notably in 2003 and during the CAP Health check in 2008, to modernise the sector and make it more market-oriented. The Europe 2020 strategy offers a new perspective. In this context, through its response to the new economic, social, environmental, climate related and technological challenges facing our society, the CAP can contribute more to developing intelligent, sustainable and inclusive growth. The CAP must also take greater account of the wealth and diversity of agriculture in the EU Member States. As a result it is adapted to meet the challenges ahead by being more efficient and contributing to a more competitive and sustainable EU agriculture. EU agriculture needs to attain higher levels of production of safe and quality food, while preserving the natural resources that agricultural productivity depends upon [14]. As shown in Figure 1, in the current situation economic, environmental and territorial challenges, sustainable use of the environment, and sustainability development are among the most important objectives.

The above information shows that in the EU common agricultural policy are given high importance such as sustainability, conservation of natural resources, safe food and quality food production issues. For these reasons, organic agriculture is to serve the basic objectives of the CAP. In this context, the development of organic agriculture is one of the important issues. Therefore, a brief information on the subject is presented below.

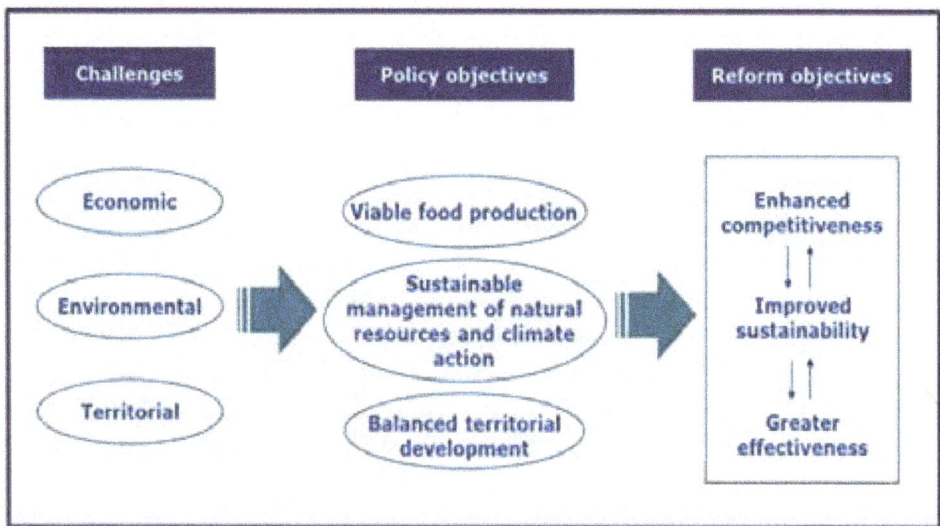

Source [14]

Figure 1. The CAP post-2013: challenges to reform objectives

In recent years, the organic movement has increasingly become a focus of policy interest in Europe. EU countries support organic farming through agri-environmental programmes and action plans, among other instruments. These policy interventions aim at both supporting consumer choice through development of the market for organic food and encouraging the provision of public goods through support for organic land management [29]. At the EU level, organic farming as a policy domain is a recent development, and arose when the CAP became more sensitive to environmental issues. In consequence, the concept of organic farming is now increasingly shaped by actors outside the organic movement.

The EU Action Plan for Organic Food and Farming was elaborated over a period of five years from a first formulation of the idea at a European Conference in Vienna 1999 to the final communication of the Commission in 2004. Organic action plans are widely used in Europe as a means of integrating different policies that can be used to support organic food and farming.

At a global level, many countries have regulatory requirements similar to those in the EU, and there are formal agreements covering trade in organic products between these countries. The International Federation of Organic Agriculture Movements (IFOAM) has formulated principles, sets international baseline standards, accredits national certification schemes to facilitate international trade and collaborates with the UN Food and Agriculture Organisation (FAO) and other international organisations to harmonise International Organic Standards [35].

2. An overview on organic agriculture

Overall for organic farming, could be called the oldest forms of agriculture on earth. But today, this definition has become quite complex. In this context organic agriculture is a holistic production management or agricultural production system that uses matter, energy knowledge and natural life for its production and processes and for providing services. Organic agriculture is a farming system which promotes and enhances agro-ecosystem health, including biodiversity, biological cycles, and soil biological activity. It emphasizes the use of management practices in preference to the use of off-farm inputs, taking into account that regional conditions require locally adapted systems [15]. This is accomplished by using, where possible, agronomic, biological, and mechanical methods, as opposed to using synthetic materials, to fulfill any specific function within the system. The production factor natural life is considered here a natural resource with particular sets of characteristics that in order to function effectively has to be respected [18]. To assess the resilience of a farming system, various elements that can build resilience are identified and the analysis shows that organic farming has a number of promising characteristics building resilience [33]. In general, the relative importance of agriculture has decreased in all European countries over last decades. However, 28 countries of the EU have significant shares in both export and demand of agricultural products in the world [8, 22]. As a result, organic agriculture is very important production system for health of human, food, soils and ecosystem.

Albert Howard is often referred as the father of the modern organic agriculture because first he applies the modern scientific knowledge and method to traditional agriculture. He worked as agriculture advisor from 1905-1924 in Bengal India where they documented the traditional Indian farming practices and came to regard them as superior to their conventional agriculture science. In Germany Rudoff Steiner's development was probably first comprehensive system of what we now call organic farming. Steiner emphasized the role of farmers in guiding and balancing the interaction of the animals, plant and soil. Healthy animal depend on healthy plant, health plant depends on healthy soil and health soil depends on healthy animal [24].

According to this, European organic agriculture emerged in 1924 based on the work of Rudolf Steiner (1861-1925) delivered his bio-dynamic agriculture course (7-16 June 1924), at Koberwitz in Silesia. His courses, the first known to have been given on organic agriculture. After Steiner, was delivered by other persons on bio-dynamic agriculture in the 1930s and 1940s. Organic agriculture was developed in Switzerland by Hans Mueller, in Britain by Eve Balfour and Albert Howard and in Japan by Masanobu Fukuoka.

In europe, numerous farms have started to convert to organic farming since 1960s, and development of organic agriculture has been supported by goverment subsidies. In many other countries, organic agriculture was adopted because of the growing demand for organic products in Europe, North America and Japan [54].

Technological advances during Second World War accelerated post-war innovation in all aspects of agriculture, resulting in large advances in mechanization, fertilization, and pesticides. In 1944, an international campaign called the Green Revolution was launched in Mexico which encouraged the development of hybrid plants, chemical controls, large-scale irrigation, and heavy mechanization in agriculture around the world. During the 1950s, sustainable agriculture was a topic of scientific interest, but research tended to concentrate on developing the new chemical approaches the one of the reasons for this, was the widespread belief that high global population growth, would soon create worldwide food shortages unless humankind rescue itself through agricultural technology. In 1962, Rachel Carson, a prominent scientist and naturalist, published Silent Spring, chronicling the effects of DDT and other pesticides on the environment [41]. In 1972, the International Federation of Organic Agriculture Movements (IFOAM) was founded and dedicated to spread principles and practices of organic agriculture throughout the world. In the 1980s, around the world, farming and consumer groups began seriously pressuring for government regulation of organic production. This led to legislation and certification standards being enacted through the 1990s and to date. Since the early 1990s, the retail market for organic farming in developed economies has been growing by about 20% annually due to increasing consumer demand.

In recent years, there has been growing trend in world trade for organic agricultural products and EU countries rank in the first place in terms of organic agricultural trade. There has been growing debate that using chemical inputs excessively following the green revolution in agricultural production raised concerns on environment and human health. For this reason, organic agriculture has developed rapidly and spread around the world [56] and organic farming is considered an alternative production sysytem. In fact, demand for organic products has increased tremendously in last a couple of decades.

IFOAM, International Federation of Organic Agriculture Movements established in 1972, a non-profit organization, is a international umbrella organisation of organic agriculture organisations, and has about 778 affiliates (members, associates, and supporters) in 117 countries [26]. The mission of this organization is to lead, unit and assist the organic movement in its full diversity and its main goal is to enhance the worldwide adoption of ecologically, socially and economically sound systems based on the principles of organic agriculture [25, 26]. As it seen, organic agriculture is not a temporary fashion since bio-dynamic farming was introduced as early as in 1924. There are also other farms of organic farming with a longstanding tradition such as organic-biological or environmentally adapted farming. Organic farming is managed in harmony with nature and the agricultural holding is mainly perceived as on organism comprising humans, flora, fauna and soil [3]. However, organic agriculture, a new model for farming development, has set the goals such as minimization of environmental pollution and sustainability of farming systems [56].

In Turkey, organic agriculture has been started in 1985 with the aiming for mainly export purposes. Turkish Association of Organic Agriculture Movements (ETO), a non-profit organization of organic agriculture, established in 1992 after twenty years IFOAM [1, 11]. By the Turkish Ministry of Food, Agriculture and Livestock, prepared the Regulation for plant and animal production based on organic production methods in 1994 [2].

According to norms of EU-Regulation on organic production of agricultural products deter-mined on June 24th 1991 the regulation (EEC) NO. 2092/91 was published in the Official Journal of the EU and became legally binding on January 1st 1993. Until then it was only valid for organic plant products. The publishing of the regulation (EEC) No 1804/99 also included organic livestock products. The regulations on livestock products became effective on August 24th 2000 [20].

Advisory or extension services have been developing for organic farming in developed and developing countries. While this services have mainly provided by producer organizations and special advisory firms in EU countries it is given by buyers of organic products products in Turkey [39].

In this study, developments of organic farming in Eropean countries and European Union members and Turkey were investigated with special attention given to advisory and extension services. More specifically, advisory service providers, how these services provided and similarities and differences were investigated between EU-28 countries, and Turkey.

3. Organic development

Evidence for significant environmental amelioration via conversion to organic agriculture is very substantial, pesticides are virtually eliminated and nutrient pollution has been substan-tially reduced loss of biodiversity, wind and water erosion, and fossil fuel use and greenhouse warming potential are all reduced in organic agriculture relative to comparable conventional agriculture systems. As it given above the agroecological characteristics of organic agriculture are reviewed-weed, invertebrate, disease, and soil fertility management practices. Yield reductions of organic agriculture systems relative to conventional agriculture average 10-15%, however these are generally compensated for by lower input costs and higher gross margins. Large-scale conversion to organic agriculture would not result in food shortages and could be accomplished with a reduction in meat consumption. Organic agriculture systems consistently outperform conventional agriculture in drought situations, out-yielding conventional agri-culture by up to 100% [30]. One argument for supporting organic farming has been that it requires more labour and leads to higher rural employment. On the other hand, the high labour costs may constrain the development of the organic sector [27, 39].

Worldwide there are 1.8 million organic producers more than three quarters of the producers are located in Asia, Africa and Latin America. The countries with most producers are India (547591), Uganda (186625) and Mexico (169570). There has been 12% increase in number of producer from 2010 to 2011. Total number of producers in 2000 were 0.25 million which and it was 1.8 in 2011, 34% are from Asia, 30% in Africa, 18% in Latin America, 16% in Europe and 1% in each Latin America and Oceania.

With regard to the growth of organic agriculture after first decade of 21st century organic agriculture area has been trebled (37.2 million hectare in 2011) as compared to 1999 (11 million hectare). According to the SOEL's Survey results, more than 37.2 million hectares are managed organically worldwide [57]. Among regions Oceania is the region with highest organic agriculture area 12.2 m ha (32,8%) followed by Europe (10.6 million ha, 28,5%), Latin America (6.9 million ha, 18,5%), Asia (3.7 million ha, 9,9%), North America (2.8 million ha, 7,5%) and Africa (1.1 million ha, 3%).

Australia (12 million ha) followed by Argentina (3.8 million ha) and USA(1.9 million ha) are the countries with highest share of organic agriculture land. The total share of top three countries is 48 percent of world organic agriculture land and 70% of world organic agriculture land is based in 10 countries with 26.3 million ha.

The world total organic agriculture land is just 0.9% of total agricultural land; again Oceania is the region where organic agriculture land is 2.9% of total agriculture land followed by Europe (2.2%) and Latin America (1.1%). China, Spain, Canada, France, Poland, Russia Federation, Kazakhstan, Turkey and Romania are the countries where the rate of increase in organic agriculture land is very high.

Much of the increase is occurring in third world countries where some farmers are attracted to the export benefits of organic food production. Many governments have been encouraging farmers to convert to organic farming for this reason, however, the study calls for a cautionary approach for the potential of export markets is often overstated. Market growth rates are slowing and supply-demand imbalances are expected to become a feature of the global organic food industry. The global market is projected to continue to expand however at slower growth rates. The industrialized world is expected to comprise most revenues, however, other regions are expected to show high growth due to the growing popularity of regional markets. The formation of trading blocs and convergence of consumer demand are also other main factors stimulating demand in other countries [54, 39].

In European Union organic farming has been implemented in almost all countries, but five of these countries – Spain, Italy, Germany, England and France-account for about 57.4% of total organic farming area. Turkey with a 325 thousands hectares organic farming area has about 3.92% organic farming area compared to EU. With respect to organic farming enterprises, the share of organic farms in Turkey is around 17% (Table 1). These indicators show that organic farming has been realized in small scale. In fact, average farm size in Turkey (9.2 ha) is significantly lower than that of EU-28 (39.7 ha).

Realizing that favorable climatic conditions and topografic structure in Turkey, organic farming area is relatively low. In order to use organic farming more effectively and increaase organic farming area in Turkey, suitable conditions need to be provided.

Country	Organic Area		Farm		Avarage
	Hectares	%	Number	%	Farm Size Scale (ha)
Spain	1330774	15.97	25291	12.05	52.6
Italy	1106684	13.28	43029	20.51	25.7
Germany	947115	11.37	21047	10.03	45.0
UK	721726	8.66	5156	2.46	140.0
France	677513	8.13	16446	7.84	41.2
Austria	518757	6.23	21000	10.01	24.7
Czech Republic	398407	4.78	2665	1.27	149.5
Sweden	391524	4.70	4816	2.30	81.3
Poland	367062	4.41	17092	8.15	21.5
Greece	326252	3.92	23665	11.28	13.8
Portugal	209090	2.51	1902	0.91	109.9
Romania	168288	2.02	3078	1.47	54.7
Finland	166171	1.99	4087	1.95	40.7
Latvia	160175	1.92	4016	1.91	39.9
Denmark	156433	1.88	2694	1.28	58.1
Slovakia	145490	1.75	363	0.17	400.8
Hungary	140292	1.68	1617	0.77	86.8
Lithuania	129055	1.55	2652	1.26	48.7
Estonia	66767	0.80	1277	0.61	52.3
Netherlands	51911	0.62	1413	0.67	36.7
Ireland	47864	0.57	1328	0.63	36.0
Belgium	41459	0.50	997	0.48	41.6
Slovenia	29388	0.35	2096	1.00	14.0
Croita	14193	0.17	885	0.42	16.0
Bulgaria	12320	0.15	379	0.18	32.5
Cyprus (South)	3816	0.05	732	0.35	5.2
Luxemburg	3614	0.04	77	0.04	46.9
Malta	26	0.00	12	0.01	2.2
Total EU-28	8307943	100.00	210501	100.00	39.7
Turkey	325 831	3.92	35 565	16.90	9.2

Source: [26, 57].

Table 1. The Avarage Scale and Number of Farm and Land Area Under Organic Management in EU and Turkey (2009)

In table 2, developments of organic farming in EU and Turkey were examined during last 14 years. Organic farming area and number of holdings main countries of the EU were showed about 5.9 and 3.5 folds increases during 1995-2009. Turkey has experienced an increase in the organic farming area (21.4 folds) but average organic farm size showed a decrease during the same period.

Countries/ Indicators	1995	2000	2004	2009	Growth rate (folds)
European Union					
Organic area (ha)	1.407.850	3.944.953	4.792.381	8332166	5.2
Number of farms	59.752	138.919	139.046	209812	3.5
Turkey					
Organic area (ha)	15.250	59.649	57.001	325.831	21.4
Number of farms	4.035	18.385	18.385	35.565	8.8

Source: [3, 13, 57].

Table 2. Development of Organic Agriculture in EU and Turkey

4. Organic product market

Worldwide organic food market has been significantly increasing. It is related to the commercial certified organic agriculture which has spread to over 130 countries worldwide. Sustained high rates of growth in sales of certified organic products in the U.S. and worldwide, averaging 20-25% in each year since 1990, have spurred concomitant growth and activities in production, processing, research, regulation and trade agreements, and exports. The global organic products market value in 2001 was estimated to be $20 billion, and the organic products share of total food sales is near 2% in the US and 1-5% in EU countries. Processed organic products have shown particularly rapid growth, often over 100% annually [30]. In 2002, the global market for organic food and drink was valued at $23 billion. Although production of organic crops is in creasing across the globe, sales are concentrated in the industrialised parts of the world. North America and Western Europe comprise the bulk of global revenues, however consumer interest is growing in other regions [39].

As an illustration of the growing international trade for organic food, its sale at retail level showed 2.5 folds increase from 9.5 billion dollars in 1996. Organic food sales mainly increased in Western countries, USA, Japan and Australia, 2.2, 2.3, 2.5 and 3 folds respectively, and 5.3 folds in other countries. These values show that demand for organic products has been growing and strong [39].

The Western European market for organic food and drink was traditionally the largest (44.3%) in the world, however it has now been overtaken by North America (37.1%). European sales of organic products were estimated to have expanded by about 8 percent in 2002 to reach $ 10.5 billion [19, 44].

Today, globally market for certified organic foods and drinks has also become fourfold larger as compared to 1999 (15billion USD) which is more than 63 billion USD in 2011 and largest market were USA 21 billion euros of organic food sales. In Europe largest market were also Germany 6.6 billion euros and France 3.8 billion euros of organic food sales. Highest per capita consumption was recorded in Switzerland (177 Euros) and Denmark (162 Euros). Denmark, Switzerland and Austria had highest shares of organic food sales [16].

It is related that organic imports into European Union members states from the developing world are growing rapidly. Certification regulations for the import of organic products are very rigorous and smallholder farmers, in common with all farmers[7]. Despite the develop-ments for organic sale in the world market, domestic market is not enough strong in Turkey and almost all of the organic food is produced for export. During the period of 2000 and 2011, organic food export in Turkey showed significant increases, rising from about 18 million to 200 million euros-an increase of nearly 11 folds. Turkey has mainly exported organic products to Germany, the Netherlands, United Kingdom, Switzerland, Northern European countries, USA, Canada, Australia and Japan are most important for Turkey. In 2023, export target of Turkish organic production is 1.7 billion Euros [1, 19].

This structure shows that Turkey has not been using its organic farming potential very efficiently. Main reasons for these can be stated as insufficient domestic demand and not using proper extension systems and approaches etc. Furthermore, poor advisory services and limited number of advisory activities are other factors that prevented the development of organic farming in Turkey [39].

5. Advisory and extension services and legislation

Organic agriculture production has different features from than conventional farming. For that reason it should be different extension and advisory services. Organic farming is realized under controlled field and applied prespecified rules. Moreover, all stages of production need to be documented. Therefore, producers and all involved firms or individuals in organic production should have enough information and experience. For this reason, all fims or individuals need to have cooperation with each others. In other words, producers are not independent as in convetional production since organic production is a controlled production system until the product reaches to consumers. There are some difficulties for organic agriculture that is mainly stem from applying internaional standards, production, processing and import of organic products, inspection procedures, labelling and marketing [39].

Especially two regulations on organic agriculture are significantly important, the US and the EU legislation which influence strongly the standards of organic production and trade worldwide. Production and inspection standards of US organic products, EU organic products and organic products from a lot of other parts of the world are equivalent with each other. The EU regulation on organic production and plant-animal products, and processed agricultural goods imported into the EU were also investigated. In the member states of the European Union (EU), the labelling of plant products as organic is governed by EU Regulation 2092/91, which came into force in 1993, while products from organically managed livestock are governed by EU Regulation1804/99, enacted in August 2000. Each European country is responsible for enforcement and for its own monitoring and inspection the system. Applica-tions, supervision and sanctions are dealt with at regional levels. At the same time, each country has the responsibility. All these regulations lay down minimum rules governing the production, processing and import of organic products, including inspection procedures, labelling and marketing [28].

As it noticed, producers, processors and export-import firms of organic products are respon-sible for certification bodies. Otherwise farmers or traders who want to export organic products should already with application for certification know the potential final destinations of their products to assure that both production standards and procedures for imported products in the aimed market are met.

In Turkey, on the other hand, the Ministry of Food Agriculture and Livestock provide extension service at the national level. In addition, due to organic production is made mostly for foreign markets, advisory and extension services for organic agricultural producers in particular are provided by private companies who contract production. Therefore, beside the Ministry, the exporter companies also provide extension service for organic farmers [9].

In Turkey, the Ministry takes a major role on extension services. According to 2006 regulation, private advisory system and advisory companies are encouraged. But the support method behind this policy is not yet enough to develop the system. Provincial directorates of the Ministry gave organic training to 1214 employees in 2003-2009 periods. In addition, they organized 2093 training programs for farmers and a total of 40.010 farmers were given training related to organic farming between 2004 and 2009. These trainings are extremely important. For this reason, the number of training activities and farmers' participation must be increased. Training activities should be more practical instead of giving theoretical information and emphasis should be placed on improving the quality and effectiveness. To achieve this, technical personnel in charge with organic agriculture departments must be provided with adequate equipment and update their professional competencies to overcome with the shortcomings in the sector. In order to develop the organic system in the central and provincial levels, a strong extension model must be developed with adequate infrastructure, physical equipment, and human resources. In order to institutionalize and build a suitable infrastructure for organic farming an effective extension model must be developed. Because as it is in developed countries, the development of organic agriculture in Turkey also depends on farmers' training and extension and advisory services provided for this purpose. Undertaking this mission in our country for many years, and organized in every province and district, the Ministry of Food Agriculture and Livestock has been carrying an important task [9].

Turkey may learn about expert knowledge and advisory systems in the EU and the roles of the stakeholders. In the European Union, advisory services are provided by the ministries, advisory companies working under ministries, agricultural chambers, private advisory firms, and private advisors. There have been an increasing trend in private advisors and advisory companies in Europe. Most of the countries want to decrease the effect of the state in advisory services, Yet, even in North-West Europe were farmers are on average much, much better off than Turkish farmers, only the most successful are prepared to fully pay their advisors. All others, including most organic farmers, rely on the state to at least contribute substantially to advisory systems. Often this is done as a project based support to research one of the many open questions concerning organic agriculture. Organic farmers in Northern Germany have therefore founded "Research and Advisory Associations" [48].

6. Advisory and extension services in Europe

Organic farming is not simple production activity for farmers becouse of they have a huge experience of conventional farming. For that reason advisory and extension service is vital for farmers. Since organic farming is applied under strict rules, producers and processors need to have high level of knowledge for organic farming. All the rules has to be performed and can be controlled by certification firms. For these reasons, extension services need to be provided by advisory units, having expert knowledge, in order to educate farmers and assure that necessary information reach to farmers at proper time. Considering the importance of advisory units, special advisory units specializing on organic farming has to be established [39].

The advisory services play an important role in transfering of scientific results into agricultural practice, ideally it should be the link between practice and research. The organic advisory services are quite well developed in the German language and in the Nordic countries which is partly integrated into the conventional advisory service. Most development in terms of advisory service is needed in the countries of southern Europe [55].

Based on development levels, countries use different extension sytems and approaches [36]. When organic farming is evaluated for farmers training and extension, fourteen major subjects should be considered:

- adopting an organic farming philosophy to farmers
- training of farmers for organic farming [39]
- training of organic farming for conventional farmers
- training of processors of organic production [39]
- teaching the benefits of organic farming to supermarket chains
- training of trainers on organic farming [39]
- training of researchers and extension worker or adviser on organic farming
- teaching the benefits of organic products to children
- making awareness of consumers on organic products [39]
- developing awareness of policy maker on organic sector
- developing awareness of all relevant direct and indirect stakeholders on organic products
- organizing promotional and educational activities (urban extension) for enhancing domestic and foreign markets
- considering the use of information technology and internet in this subject [39]
- developing awareness to cooperate all relevant public, farmers' organization, and private sector on organic products.

One of the major aims of this study is to examine advisory services for organic agriculture in EU member states and Turkey. These advisory units were examined in four groups based on services provided to farmers such as fundemantal knowledge required for organic farming, increasing problem solving ability of farmers, deciding suitable product variety and purchasing production inputs. These four groups were determined as Private advisory services, Farmer's organizations, Ministries of agriculture and Others. In table 3, advisory service units in EU and some other European countries and Turkey were presented. All countries have more than one or more advisory units having different characteristics.

Advisory Service Units	TOTAL	Switzerland	Serbia	Norway	Macedonia	Albania	Total (EU-28)	Total (EU-13)	Croita	Malta	Cyprus (south)	Estonia	Hungary	Lithuanixa	Latvia	Czech Rep.	Bulgaria	Slovakia	Slovenia	Romania	Poland	Total (EU-15)	U.K.	Sweden	Spain	Portugal	Netherlands	Luxemburg	Italy	Ireland	France	Greece	Germany	Finland	Denmark	Belgium	Austria	Turkey
Private Advisory	3						26	12														14																
Local advisors direct contact with farmers	-						1	1														1														X		
Private advisory institutes(ökoringe)	-						2	2														2											X					
Private consulting advisory firms	1	X	X				10	7				X	X				X	X	X	X	X	3	X		X				X			X	X					
Companies selling input / commercial advice/ Input producers	-						4	2									X					2				X						X						
Private/Exporter firms via contract farming	1						-	-														-				X					X	X						X
Conventional advisory firms	-						3	-						X								3	X	X			X				X							
No state advisory service	1					X	2	1														1	X											X				
Independent advisory/consultant	1		X		X		4	2								X				X		2	X			X		X				X						
Farmer's Organizations	2	X			X		19	7			X	X			X					X	X	12	X		X				X		X	X	X				X	
Organic Producer organizations	2						7	4														3										X						
Producer organizations/Chamber of Agric.	-						4	1		X											X	3								X			X				X	
Organic association advisors	-						3	1														2	X	X	X		X	X		X								
Organic agriculture consulting units	1				X		5	1						X		X					X	4	X	X		X	X	X				X						
Ministries/Public	6						19	13														6															X	
Ministry of Agriculture	2		X				4	4		X	X	X	X			X		X				-															X	X
Min.of Agric. in organic farm unit	-						2	-														2												X				X
Regional agric.authorities(in districts)	2		X	X			3	1														2	X						X				X					
Min.of Agriculture, organic conversion information service	-						2	1						X						X		1																
National Agric.Advisory Service	1			X			2	1				X			X	X		X	X		X	1													X			
Ministry of Agriculture& other Organization/institute/centre	1						6	6	X						X	X		X				-																X
Others	1						-	-														7																
Agr.schools/courses/seminar/demonstration	1	X	X		X		5	-		X		X						X			X	5		X				X				X					X	X
Exchange between organic farmers	-						1	-														1										X						
Telephone helpline	-						1	-														1	X															
Advisory service units number	12	1	2	2	3	1	71	32	2	2	2	3	1	2	2	4	2	3	2	4	4	39	5	3	3	2	2	2	2	2	2	4	4	1	2	4	4	3

(*) Based on author's computations from [4, 5, 6, 10, 17, 21, 23, 31, 32, 34, 36, 37, 40, 42, 43, 45, 46, 47, 49, 50, 51, 52, 53, 55, 58].

Table 3. Advisory Services for Organic Agriculture in Europen Countries (*)

In order to see the number and share of these units, table 4 was produced from the table 3. Similar units according to management and functions were considered in the same group.

| Advisory Service | The number of advisory unit service (European Union) | | | | | | The number of advisory unit service Europe (34 Countries) | |
| | EU-15 (a) | | EU-13 (b) | | EU-28 (a+b) | | | |
	Number	Share (%)	Number	Share (%)	Number	Share (%)	Number	Share (%)
Private services	14	35.9	12	37.5	26	36.6	29	34.9
Farmer's organization	12	30.8	7	21.9	19	26.8	21	25.3
Ministry of agriculture	6	15.4	13	40.6	19	26.8	25	30.1
Others	7	17.9	0	0.0	7	9.9	8	9.6
Total	39	100.0	32	100.0	71	100.0	83	100.0

Table 4. Distribution of Advisory Units in European Countries

As shown in table 4, the most important advisory units in EU-15 are private advisory services (35.9%) and farmer's organizations (30,8%). These units are followed by the ministry of agriculture (15.4%) and others (17.9%). But the situation for the EU-13 is different, the main advisory units are private advisory services (37,5%) and ministry of agriculture (40,6%). These units are followed by farmer's organizations (21,9%). When all EU Members and non-member European countries are considered together, the most important advisory units are private advisory services (34.9%) and ministry of agriculture (30,1%). These units are followed by farmer's organizations (25.3%) and others (9.6%). As can be seen in tables 3 and 4, there are some differences between the first 15 members of the European Union (EU-15) and the 13 countries. These differences are related with their extension systems. Especially after 2004, the higher presence of the ministry of agriculture was interesting in the farmer's organization.

Generally, in the group of private advisory, the most important services provided are called Private consulting advisory firms and conventional advisory firms, however, Organic agriculture consulting units and Organic Producer organizations are important services in the farmer's organizations. In the ministries, the most important services are Ministry of Agriculture in organic farm unit and Regional agriculture authorities in districts.

Since organic farming is mainly considered for export purposes, advisory services for organic products have been implemented by Private firms via contract farming (foreign exporter firms) which is the most important advisory unit for contract farmers. For this reason, advisory units have been mainly formed and managed by importing firms in Turkey.

In addition to this, Ministry of Food Agriculture and Livestock has been conducting advisory services for organic agriculture with cooperation of Universities and Turkish Association of Organic Agriculture Movements (ETO). It is expected that these units will have significant importance in providing extension services in Turkey. However, the major difference bettween EU-15 countries and Turkey is the lack of producer organizations in Turkey.

7. The public extension model for organic agriculture of Turkey

In Turkey, mostly extension and advisery services for organic agriculture are implemented by public (the Ministry of Food Agriculture and Livestock). In order to institutionalize and build a suitable infrastructure for organic farming an effective and applicable extension model must be developed for the country. Because as it is in developed countries, the development of organic agriculture in Turkey also depends of farmers' training and extension and advisory services provided for this purpose. Undertaking this mission in our country for many years, and organized in every province and district, the Ministry of Food Agriculture and Livestock has been carrying an important task and the organic farming units of the Ministry are carrying out their mission.

Currently organic agriculture units in 81 provinces mostly employed in Farmer Training and Extension Division are serving as extension agents. In general, except a couple of provinces, 2-5 personnel are working in organic farming units. These aren't completely devoting their efforts for extension activities but for some other departmental work. Considering provincial potential of organic farming and personnel status of each province, the number of personnel working for organic units must be arranged. The determined leader provinces will also make contributions to the neighboring provinces due to similarities in environmental conditions and agricultural structures [9]. Therefore, these provinces can be identified as "Organic Farming Central Provinces" (OFCP). Administratively no province will supersede each other but the provincial directorates will work in a cooperative way with a professional manner (Figure 2). Once the professional staff members of OFCP work effectively, they will probably develop organic farming in their own province and will make further contributions to neighboring cities. In this way professional staff of organic farming units working in provinces with lower and/or uneconomical organic potential may concentrate on other activities. In addition, the TARGEL (public extension) personnel must also be used for organic agriculture especially in villages with high organic potential. All this process will increase the effectiveness of extension services and will contribute to organic farming as well as proper use of human resources in the Ministry.

In the new suggested extension advisory system, the following measures must be taken to make the personnel of OFCP and OFU work effectively [9, 38]:

- *Ensuring full-time work in organic agricultural extension:* Among the other duties and responsibilities, personnel working in organic farming must perform two major tasks; control and advisory. However, according to agreement with the international sense, control and extension duties and responsibilities should not be undertaken by the same personnel. The control function is completely different from the advisory function. In this case, the extension advisor and the controller must be different people and independent of each other. Accordingly, workload of the personnel working in the field of organic agriculture must be reduced.

- *Technical information capacity must be increased:* In every region there is a need of expertise and technical information in organic farming activities. Therefore, personnel working for

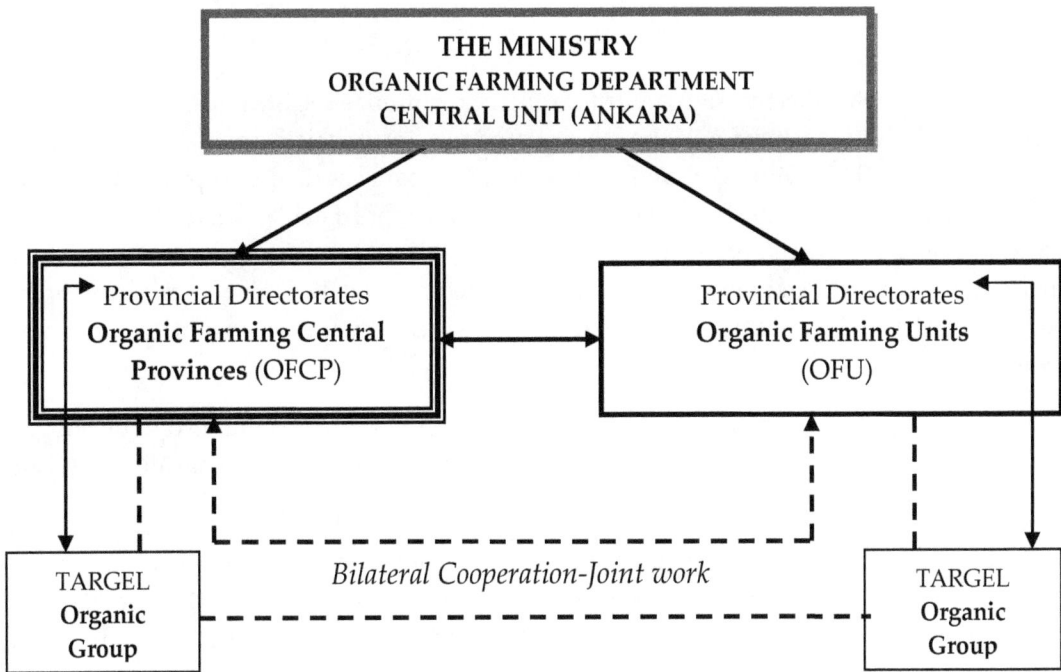

Figure 2. Re-organizing public extension for organic farming in Turkey [38].

the extension services of organic farming must be provided with technical information and update their professional competencies.

- *Personnel must be gained practical skills and field experience*: Beside theoretical information, technical staff must be trained in the field to gain practical experiences, particularly for the specific region. They also must participate in hands-on training activities and be able to apply their skills in the field.

- *Adoption of participatory extension approaches:* It is important to utilize the positive sides of the participatory extension approaches. These approaches require institutional changes and appropriate arrangements for the implementation. In this way, participatory approaches must be taught to extension personnel and the efficiency of extension work must be increased.

- *Continuous in-service education to update and upgrade occupational knowledge of extension personnel:* In order to develop well-educated and skillful extension personnel extension staff must be given in-service education in certain times. In order for the extension personnel to use proper extension methods and to give farmers correct information them must take a "training of trainers" program. This can be possible with adoption of continuous education programs [9, 38].

As it can be seen, if the public wants to provide effective extension or advisory services for organic agriculture the most appropriate model should be develop or use in the actual conditions in Turkey.

8. Conclusions

As a result, the main objective of organic farming is to optimize the life circle and to establish a sustainable farming system for the health and productivity of societies of plants, animals and people. Accordingly healthy food and in this context organic agricultural products are vital for all nations. Communities are willing to consume more healthy foods that has been controlled by government or related organizations. Nowadays the quality of food is an emerging trend for consumers especially in developed nations. For this reason, farmers as a food producers are required to benefit from the education and extension services more frequently.

Demand for organic food has been increasing not only in developed but also in developing countries in the last two or three decades. Due to characteristics of organic farming, advisory units has to be establish at the production decision.

Because of high price, the demand for organic food is not expected to increase in a short period of time in Turkey. However, increasing awareness of consumers to health concern might stimulate the demand for organic products in the long term. Regarding advisory units, the approaches used in the EU countries might be used in Turkey such as establishing farmers and private organizations [39].

The contract farming approach which was initiated by importing firms must be considered for Turkey's organic sector. Furthermore, there is a urgent need to stress on encouraging advisory units to achieve the competitiveness and efficiency.

Both, the EU and other countries, it is necessary to benefit from the experiences. Establishing a network to share experiences would be useful to every country.

Participatory extension approaches which are very effective, would be very helpful in the use of organic agricultural producers. Farmers' organizations can play an important role to increase the efforts unfortunately these organizations are not effective. Countries should contribute to the development of organic farming because domestic demand for organic agricultural products has been increasing day by day. Each country according to their own conditions to establish the most effective advisory system.

Author details

Orhan Özçatalbaş[*]

Address all correspondence to: ozcatalbas@akdeniz.edu.tr

Akdeniz University, Faculty of Agriculture, Department of Agricultural Economics, Antalya, Turkey

References

[1] Aksoy,U., 1999. Organic Agriculture in the World and Turkey. First Symposium on Organic Agriculture. 21-23 June 1999, Ege University, Izmir.

[2] Anonymous, 1994. The Regulation for Plant and Animal Production Based on Organic Production Methods. 18 December 1994, No: 22145 Official Journal of Republic of Turkey, Ankara.

[3] Anonymous, 2001. Organic Farming in Europe-Provincial Statistics 2001 (.http:// organic-europe.net/ europe_eu/statistic.asp). SÖL:Stiftung Ökologie & Landbau, Germany.

[4] Anonymous, 2011. Liechtenstein-Country Report. FiBL Sippo 'The Organic Market in Europe Publication, Switzerland. (www.organic-europe.net/country-report-liectenstein.html)

[5] Anonymous, 2013. Organic agriculture Serbia at a glance 2013. FiBL 2013, Switzerland.   (http://www.organic-world.net/news-organic-world.html?&L=0&tx_ttnews %5Btt_news %5D=1049&cHash=a39f8ab896e47a8e2087d2dd2326d5a8).

[6] Apostolov, S., 2012. Bulgaria: Country Report, The World Of Organic Agriculture. (www.organic-europe.net/1684.html)

[7] Barrett, H.R., A.W. Browne, P.J.C. Haris, K. Cadoret. 2001. Smallholder Farmers and Organic Certification: Accessing the EU Market from the Developing World. Biological Agriculture and Horticulture, Vol. 19 (2) 183-199,

[8] BML, 1999. Agriculture and Forestry in Germany. Data and Facts 1999. The Federal Ministry of Food, Agriculture and Forestry.Printed by Mintzel-Druck, 95028. Hof/ Saale. feb.1999,Germany.

[9] Boz, İ, Aksoy, U., Özçatalbaş,O., Enhancing Organic Farming in Turkey. Workshops and Training Programmes, UTF/TUR/052/TUR, July 2011, Published by FAO.

[10] Dezseny, Z., and Drexler. D., 2013. Organic Agriculture İn Hungary-Past, Present and Future, FİBL and IFOAM (2013) The World of Organic Agriculture, Frick and Bonn. (www.organic-europe.net/2547.html)

[11] Ertem, A., 1993. Organic Agriculture and Rapunzel, Izmir.

[12] ETO,2012, Organik Tarım Ürünleri İhracatında İşbirliği bildirisi(Organic Agricultural Products Export Cooperation), Turkish Ecological Agriculture Organization Association (Ekoloji Tarım Organizasyonu Derneği), İzmir.

[13] EU, 2012. The Common Agricultural Policy. European Commission Agricultural European Union. ISBN: 978-92-79-23265-7. printed in Belgium

[14] EU, 2013. Overview of CAP Reform 2014-2020, European Commission Agricultural Policy Perspectives Brief. No 5, December 2013 (http://ec.europa.eu/agriculture/policy-perspectives/policy-briefs/05_en.pdf)

[15] FAO, 1999. Food Standard Programme Codex Alimentarius Commission Report Joint FAO/WHO. Rome.

[16] FiBL, 2011. The World of Organic Agriculture, Statistics and Emerging Trends, 2011. FiBL & IFOAM. Switzerland.

[17] Frieden, C., and Huber, 2010.Country Report About Organic Agricultur in İceland. FİBL-Sippo 'The Organic Market İn Europe', Handbook. Switzerland. (www.organic-europe.net/country-report-iceland.html)

[18] Goewie E.A., 2002. Organic Agriculture in the Netherlands; Developments and Challenges. Netherlands Journal of Agricultural Science, Vol:50 (2): 153-169.Royal Netherlands Soc. Agr. Sci, Wageningen.

[19] GTHB, 2013, Organic Agricultural Products Exports and Imports. Republic of Turkey Ministry of Food, Agriculture and Livestock, (http://www.tarim.gov.tr/). Ankara.

[20] GTZ, 2004. Organic Agriculture. (http://www.gtz.de/organic-agriculture/ english/ com/com. html) Postfach 5180 D-65726 Eschborn, Germany.

[21] Guda, A., 2009. Country Report About Organic Agriculture in Albania. Sasa Project, Tirana. FİBL, Switzerland. (www.organic-europe.net/country-info-albania-report.htmı)

[22] Haccius, M., I.Lünzer, 2000. Organic Agriculture in Germany. (.http://organic-europe.net, 23.06.2000). Stiftung Ökologie & Landbau(SÖL), Bad Dürkeim-Germany.

[23] Heinonen, S., 2009. Organic Agriculture in Finland 2008. Fibl 2012. Switzerland. (www.organic-europe.net/finland.html)

[24] Holger Kirchmann; Gudni Thorvaldsson, Lars Bergström, Martin Gerzabek, Olof Andrén, Lars-Olov Eriksson and Mikael Winninge (2008). Holger Kirchmann and Lars Bergström, ed. Organic Crop Production – Ambitions and Limitations. Berlin: Springer. pp. 13–37.

[25] IFOAM, 2004. Uniting the organic world. International Federation of Organic Agriculture Movements (http://www.ifoam.org/).

[26] IFOAM, 2013. The Organic Movement Worldwide: Directory of IFOAM Affiliates 2013 International Federation of Organic Agriculture Movements (http://www.ifoam.org/en/about-us-1).

[27] Jansen, K., 2000. Labour, livelihoods and the quality of life in organic agriculture in Europe. *Biological Agriculture & Horticulture*, 17 (3): 247-278 2000.

[28] Kilcher, L., B. Huber, O. Schmid, 2004. Standards and Regulations. The World of Organic Agriculture-Statistics and Emerging Trends-2004. (eds:Willer,H., M.Yussefi) IFOAM, Bonn,Germany.

[29] Lampkin, N. H. and M. Stolze (2006). EuropeanAction Plan for Organic. Food and Farming. Law, Science and Policy. 3: 59-73.

[30] Lotter, D.W., 2003. Organic agriculture. *Journal of Sustainable Agriculture*, 21 (4): 59-128 2003

[31] Maciejczak, M. And Matera, D., 2010. Food Safety in the New Member States. IUCN programme Office for central Europe. Brussels, Belgium.

[32] Matera, D., 2005. Organic Agriculture in Poland. FiBL 2012, Switzerland, (http://www.organic-europe.net/fileadmin/documents/country_information/ARCHIVE/poland-2005-organic-europe.pdf)

[33] Milestad, R., and Darnhofer, K., 2003. Building farm resilience: The prospects and challenges of organic farming. *Journal of Sustainable Agriculture*, 22 (3): 81-97 2003

[34] Moschitz H., and M. Stolze, 2007. Policy networks of organic farming in Europe. Organic Farming in Europe:Economics and Policy, Volume 12. Published by Universitat Hohenheim, Stuttgart, Germany..

[35] ORC, 2013. Overview of European Regulation on organic food. The Organic Research Centre, Elm Farm, Hamstead Marshall, Newbury, Berkshire RG20 0HR, UK. (http://www.organicresearchcentre.com/).

[36] Özçatalbaş, O., Y. Gurgen, 1998. Agricultural Extension and Communication. Baki Publication. ISBN: 975-72024-02-3, Adana.

[37] Özçatalbaş, O., 2000. Horticultural Information System and Extension Organization in Hannover Region, Germany. Hannover University, Horticultural Faculty, Institute of Horticultural Economics. Hanover, Germany (unpublished report).

[38] Özçatalbaş, 2010. Türkiye'de Organik Tarımın Güçlendirilmesi Bölge için Yeni Fırsatlar:Fındık ve Diğer Ürünler Çalıştayı Raporu, Samsun.

[39] Özçatalbaş,O., R.Brumfield, B.Karaturhan,2010. Advisory services for organic agriculture in the European Union and Turkey.Journal of Food, Agriculture & Environment Vol.8 (2) : 507-511. 2010, WFLPublisher Science and Technology Meri-Rastilantie 3 B, FI-00980 Helsinki, Finland.

[40] Padel, S., 2001. Information and Advisory Services for Organic Farming in Europe. 15[th]ESEE Integrating Multiple Landuse for a Sustainable Future Seminar. August 27-31, 2001. Wageningen-The Netherlands.

[41] Paull, John 2007. "Rachel Carson, A Voice for Organics-the First Hundred Years". Journal of Bio-Dynamics Tasmania. pp. (86) 37–41.)

[42] Prokopchuk, N. and Eisenring, T., 2011. Switzerland Country Report. Research Institute of Organic Agriculture (FiBL) Switzerland, (www.organic-europe.net/country-info-ukraine-report.html?&L=0#c6433).

[43] Radulovic, J., 2008. Country Report About Organic Agriculture In Montenegro 2008, FİBL, Switzerland. (www.organic-europe.net/country-info-montenegro.html)

[44] Sahota, A., 2004. Overview of the Global Mar ket for Organic Food and Drink. The World of Organic Agriculture-Statistics and Emerging Trends-2004. (eds:Willer,H.,M.Yussefi) IFOAM, Bonn,Germany.

[45] Slabe, A,. 2012. Organic Farming in Slovenia. Institute for Sustainable Development (ISD) Slovenia, BiBL 2012 Switzerland. (http://www.organic-europe.net/slovenia.html?&L=0)

[46] Stefanescu, S.L., 2012. Current Status of the Agricultural Advisory and Extension System in Romania. Education-Extension Coordinator, PMU MAKIS/CESAR, Bucharest, Romania.

[47] Stefanescu, S.L., Steriu,S. and Dumitraşcu, M., 2013. Private and Public Players on the Market Agricultural Advise and Extension in Romania. 21st European Seminar on Extension Education: Extension education worldwide, September 2-6, Antalya, Turkiye

[48] Thimm, C., et al, 2011. Enhancing Organic Farming in Turkey, FAO, UTF/TUR/052/TUR Sub-regional Office for Central Asia, February 2011. Published by FAO.

[49] Trajkovic, R., 2011. Organic Farming in The Farmer Yugoslav Republic of Macedonia. FİBL, Switzerland. (www.organic-europe.net/country-info-macedonia-fyrom.html)

[50] Urban, J., 2012. Organic Agriculture in The Czech Republic (Country Report 2011). In: Wiesinger, K., and Cais, K., (eds) Bayerische Landesanstalt Für Landwirtshnft, Freising, No. 4/2012 pp.169-178. (http://orgprints.org/20991/)

[51] Vatemaa, A., and Milkk, M., 2009. Organic Farming in Estonia. FİBL, Switzerland. (www.organic-europe.net/estonia.html)

[52] Willer, H. and Niggli, U., 2010. Switzerland-country report. FiBL Switzerland, (http://www.organic-europe.net/switzerland.html?&L=0#c929).

[53] Willer, H., I.Luenzer, M.Haccius, 2002. Organic Agriculture in Germany 2002, (Update: October 2002 http://www.organic-europe.net/country_reports/default.asp).

[54] Willer, H., M.Yussefi, 2001. Organic Agriculture Worlwide 2001, Statistics and Future Prospects. Stiftung Ökologie & Landbau –SÖL No:74. Bad Dürkeim-Germany.

[55] Willer, H., U. Zerger, 1999. Demand of Research and Development in Organic Farming in Europe. FAO Worksop on Research Methodologies in Organic Agriculture at

the FIBL(.http://wwwsoel.de/inhalte/oekolandbau/research_intro.html).Frick, Switzerland.

[56] Xie, B., X.R.Wang, 2003. Organic Agriculture in China, *Journal of Outlook on Agriculture*. Vol.32 (3): 161-164.

[57] Yussefi, M., 2004. Development and State of Organic Agriculture Worldwide. The World of Organic Agriculture-Statistics and Emerging Trends-2004. (eds:Willer,H., M.Yussefi) IFOAM, Bonn,Germany.

[58] Zarina, L., 2009. Organic Farming in Latvia:Country Report. FiBL, Switzerland. (www.organic-europe.net/latvia.html)

3

Application of Active EM-Calcium in Green Agricultural Production — Case Study in Tomato and Flue-cured Tobacco Production

Xiaohou Shao, Tingting Chang, Maomao Hou,
Yalu Shao and Jingnan Chen

1. Introduction

EM (Effective Microorganisms) technology has long been widely applied to agricultural production in China. EM contains more than 80 kinds of beneficial microorganisms, including yeast, lactic acid bacteria, actinomycetes and photosynthetic bacteria (Shao et al., 2013; Zhou et al., 2008). They can quickly decompose organic matter, metabolize antioxidant substances and inhibit the proliferation of harmful microorganisms (Higa, 1997). EM was supposed to have positive effects on reforming soil nematode community structure (Hu and Qi, 2013a), increasing crop yield and improving soil properties (Daly and Stewart, 1999; Khaliq et al., 2006).

High calcium content in plants has been associated with increased resistance to diseases (Berry et al., 1988), not merely the well-known tomato blossom-end rot (BER) (Sonneveld and Voogt, 2009), calcium nutrition significantly affects the resistance of tomato seedlings to the bacterial wilt caused by Ralstonia solanacearum (Yamazaki, 1995), other diseases were also reported to be suppressed by high calcium concentration in host plants (Almeida et al., 2009; Berry et al., 1988; Chiasson et al., 2005). BER is a major physiological disorder in tomato that creates up to 50% losses (Taylor, 2004), besides the inducing factors of high temperature, high salinity (Adams, 1993), BER has long been considered to be a Ca-deficiency disorder, and its incidence increases in cultivation at low Ca concentrations (Raleigh, 1944), however, most tomatoes contain little calcium (Ca^{2+}), uptake and translocation of cationic nutrients including Ca^{2+} in plants plays an essential role in the physiological processes (Chung et al., 2010). According to early studies, the exogenous calcium supplied through tomato roots (Hall, 1977; Sachan and Sharma, 1981) or through leaves (Eraslan et al., 2007; Freitas et al., 2012) by the spraying methods can both increase the Ca uptake of tomato plants and enhance the immunity to diseases.

Flue-cured tobaccos are important industrial crop significant to the national economy for China. In southwest China, most of the rain falls during the early and middle growth stages of flue-cured tobacco, and periodic drought often happens at the later growth period of flue-cured tobacco, this will not enable the upper leaves maturing normally, resulting in a poor availability of upper leaves, so it is important to discover technologies for the drought resisting and the tobacco quality and yield improvement (Hou et al., 2012; Hou et al, 2013). Water-retaining agent has been applied in many areas (Kumar and Dey, 2011; Truax and Gagnon, 1993). However, in studies of flue-cured tobacco cultivation, application of the water-retaining agent is still in a research vacant. At present, many studies have reported the application of EM in agricultural production (Hu and Qi, 2013a, b; Javaid, 2010; Khaliq et al., 2006). EM can quickly decompose organic matter, metabolize antioxidant substances and inhibit the proliferation of harmful microorganisms (Daly and Stewart, 1999; Daming, 1999; Heo et al., 2008).

In conclusion, many studies (Heo et al., 2008; Hu and Qi, 2013b; Khaliq et al., 2006) have been undertaken on the application of EM as base fertilizer or as a component of base fertilizer, however, there has been little published concerning using EM as foliage fertilizer, especially making EM into calcium nutrient solution. In recent years, our team invents a new product based on EM technology and names it as Active EM-Calcium. Active EM-Calcium is prepared by special fermentation process with lime and EM, which contains chelating calcium (Ca) and microorganisms in the solution. In order to verify the effect of Active EM-Calcium on tomato and flue-cured tobacco production, two researches were carried out from 2012 to 2013.

2. Effects of EM-Calcium on production of greenhouse tomato

2.1. Materials and methods

2.1.1. Test site

The greenhouse experiments were carried out in plastic greenhouse during 2012 at the Vegetables and Flowers Institute of Nanjing (SW of China, lat. 31°43' N, long. 118°46' E) in a well stirred heavy lay loam. The local climate is subtropical monsoon, with average annual rainfall of 1106.5mm, temperature of 15.7°C and humidity of 81%. Affected by the No. 1211 (international numbering) severe tropical storm "HAIKUI", temperature of experimental field in early August saw a drop of 10°C-14°C, mixed with heavy winds and rains, but due to the short continuance, greenhouse crops were little influenced by the storm. Average temperature during the experiment period was 21.9°C (May), 25.7°C (June), 29.8°C (July), 27.9°C (August), 22.3°C (September) respectively, which was 1.0°C, 0.9°C, 1.5°C,

$0.6°C,-1.0°C$ higher than the same period in recent years. Peak temperature was $37.9°C$ recording on July 29th. The experimental soil properties were as follows: organic matter 14.34 g kg[-1], available nitrogen 104.17 ppm, available phosphorus 26.48 ppm, available potassium 184.70 ppm.

2.1.2. Experimental design

Lab experiment: 28 g lime was accurately weighed in a 100 ml volumetric flask, then added DI water to dissolve and kept constant volume to 1000 ml, and the 2% Ca^{2+}suspension was available by shaking the mixture well. Then the prepared calcium suspension was mixed with different volume EM, molasses (EM and the molasses were supported by EMRO Limited Company, Nanjing Branch) and DI water to make up 5 treatments with three replications. The ingredients of treatment T1-T5 were displayed as Table 1. These sealed EM-Calcium mixtures would be fermented for 3-6 days in a orbital shaker with constant temperature, they were taken out and setted aside when the precipitates were basically dissolved. The fermented treatment with highest calcium solubility was diluted to keep the active EM-calcium solution with 2.0‰ Ca^{2+}concentration.

Treatment	2% Ca^{2+} suspension (ml)	EM (ml)	Molasses (ml)	DI water (ml)	Theoretical Ca^{2+} concentration
T1	45	25	25	205	3.0‰
T2	60	30	30	180	4.0‰
T3	75	40	40	145	5.0‰
T4	90	45	45	120	6.0‰
T5	105	50	50	95	7.0‰

Table 1. The components of EM-calcium solution with different contents.

Field experiment: the six week old tomato seedlings ("*21st Century Crown*") were transplanted to the experiment fields on June 7th, conventional field management were carried out fairly among the treatments, no additional light, heat, or CO_2 were provided. The experimental field was ploughed several times and fertilized with 700kg hm[-2] compound fertilizer (N: P_2O_5: K_2O=1:2:2) in May. Irrigation systems for tomato were accorded with the local farming practices. At 8:30 every morning, micro sprayers were adopted to spray the active EM-calcium solutions 2 ml once in four days on different tomato organs. The treatments were set based on the sprayed organs, including spraying root, spraying flower, spraying leaves near the newborn fruits, spraying one week old fruit, spraying three week old fruit, and a control, hereafter referred as SR, SF, SL, SO, ST and CK respectively. A plastic film was used to keep apart the other parts when spraying one tomato organ.

2.1.3. Measurements

When tomato was ripened, two fruits were harvested from one plant, and about 10 g tomato flesh per fruit was taken along the longitudinal axis (24 fruits per treatment in total) then homogenized for the following measurements. Different forms of Ca in the tomato fruits were extracted and determined by adopting Ohat Y's method (Ohat Y, 1970), including Ca nitrate and Ca chloride, water soluble organic Ca, Ca pectate, Ca phosphate and Ca carbonate, Ca oxalate, Ca silicate, hereafter the above Ca forms were recording successively as Alc-Ca, H_2O-Ca, NaCl-Ca, HAC-Ca, HCl-Ca, Res-Ca based on the extraction solvent type. Other quality indexes were evaluated by common testing methods (AOAC, 1990; Wang et al., 2011; Yang et al., 2012): Vitamin C content was measured by the 2, 6-dichloroindophenol titrimetric method; soluble sugar was measured by the anthrone method; soluble protein was measured by the Coomassie brilliant blue method; nitrate content was measured by the ultraviolet spectropho-tometry method. For each treatment, thirty tomato fruits with a red or orange color were collected randomly to determine the basic morphological parameters, containing long diameter (L-diameter) and short diameter (S-diameter). Fruit weight was calculated from the total number and weight of fruits harvested.

2.2. Results

2.2.1. Calcium solubility in LAB experiment preparation of EM-Calcium mixtures

According to the survey results from the measured Ca^{2+}concentration, the Ca solubility was calculated and showed in Table 2. Results indicated that the measured Ca^{2+}concentration of T5 and T4 was significantly higher (P<0.05) than that of T1, T2, T3. The Ca solubility of T4 reached the peak value of 89.5%, which was 22.17%, 31.00%, 9.90%, 9.79% higher than that of T1, T2, T3, and T5 respectively. Therefore T4 was selected as the optimum formula applying to the following field experiment on account of high calcium solubility. The eventually diluted T4 solution used for the field experiment was acid, with the pH value of 4.7.

Treatment	Theoretical Ca²⁺ concentration	Measured Ca²⁺ concentration	Solubility
T1	3.0‰	2.02‰od	67.33%
T2	4.0‰	2.34‰oc	58.50%
T3	5.0‰	3.98‰ob	79.60%
T4	6.0‰	5.37‰oab	89.50%
T5	7.0‰	5.58‰oa	79.71%

Means within a column followed by the same letter are not significant at P<0.05 (Duncan's multiple range test).

Table 2. Theoretical and measured Ca²⁺concentration after 6 days fermentation

2.2.2. Forms of Ca in fruit

The Ca accumulation in tomato fruits can directly reflect the effect of EM-calcium solution. Fig. 1 gave the content of different forms of Ca in tomato fruits with different treatments. According to the survey results, the EM-calcium application was mainly beneficial for the accumulation of Alc-Ca, H_2O-Ca, NaCl-Ca, and HAC-Ca, and which had no obvious effects on the HCl-Ca and Res-Ca accumulation.

Ca pectate (NaCl-Ca) was the main Ca form presented in the ripe tomato fruits, occupied more than 75% of total Ca content, as the figures shown. The NaCl-Ca content of SR, SF, SL, SO was significantly higher ($P<0.05$) than that of CK, and the increases of SL were the most obvious, closely followed by SO, which implied that SL and SO were conductive to the accumulation of NaCl-Ca. While no significant difference ($P>0.05$) for NaCl-Ca content was observed between ST and CK. Water-soluble calcium The EM-calcium application also had great influences on the accumulation of water-soluble Ca including the Alc-Ca and H_2O-Ca in tomato fruits. Based on the results, SL increased the content of Alc-Ca and H_2O-Ca most significantly ($P<0.05$) by 2.19 and 0.71 multiples respectively in comparison with CK. Besides, SR, SF, SO and ST increased Alc-Ca content significantly ($P<0.05$), and SR, SF increased H_2O-Ca content significantly ($P<0.05$), compared to CK. The H_2O-Ca content in the tomato fruits of SO and ST was significantly different ($P<0.05$), but there were no significant differences between that of SO and CK, either ST and CK.

HAC-Ca content of SL and SO was significantly higher ($P<0.05$) than that of CK by 65.68% and 36.58% respectively, and the differences among CK, SR, SF and ST were not significant ($P>0.05$). Dissimilarly, HCl-Ca content in SF, SL, SO was 24.77%, 29.05%, 10.05% lower than that in CK, and the effects of ST and SR on the variation of HCl-Ca content were not obvious. Res-Ca accounted for the smallest share of total Ca, except for ST, no significant differences of which were found among the treatments ($P>0.05$), indicating that EM-calcium application had little effects on the Res-Ca accumulation in tomato fruits.

2.2.3. Fruit quality

Fig. 2 showed some quality indexes of tomatoes observed with different spraying methods. Vitamin C (L-ascorbic acid) is essential for all living plants where it functions as the main hydrosoluble antioxidant (Lima-Silva et al., 2012). SF, SL, SO increased the vitamin C of tomato fruit significantly ($P<0.05$) compared to CK, the increases of SL was most obvious,with the value of 25.38%. However, SR and ST did not affect the content of vitamin C greatly.

Tomato taste quality is largely determined by the contents of soluble sugar (Dorais, 2001). According to the results, the soluble sugar content of SL and SO was significantly higher ($P<0.05$) than that of CK by 9.65% and 7.20% respectively, but there were no significant differences ($P>0.05$) among that of CK, SR, SF and ST.

SL obtained the highest content of soluble protein, which was significantly higher ($P<0.05$) than that in CK. SO took the second place, slightly lower than SL but no significant differences

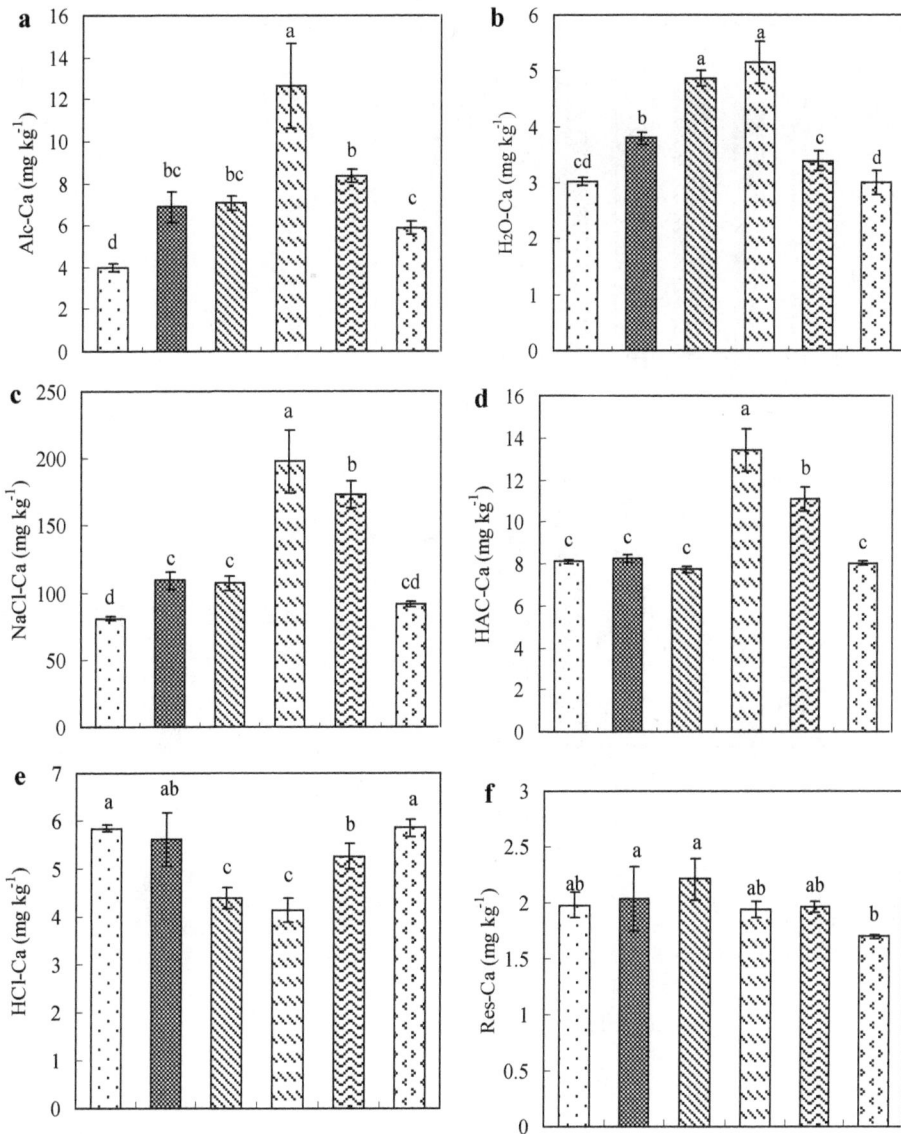

□ CK ▓ Spraying root ▨ Spraying flower ▫ Spraying leaves ▨ Spraying one week old fruit ▱ Spraying three week old fruit

Figure 1. The accumulation of different forms of Ca in tomato fruits after the application of EM-calcium solution on different tomato organs. Columns with the same letter represent values that are not significantly different at the 0.05 level of probability according to the Duncan's multiple range test. Each value is the mean ± SD (n=3). The treatment symbols are the same as the experiment design.

(P>0.05) were found between them. SF and ST had little effects on the content of soluble protein in tomato fruits. Compared to CK, the nitrate content in the other treatments was significantly lower (P<0.05), and SL was particularly apparent. In terms of these basic quality indexes of tomato fruits measured, SL was suggested as the preferable treatment since the above quality indexes of SL were all at the satisfactory levels.

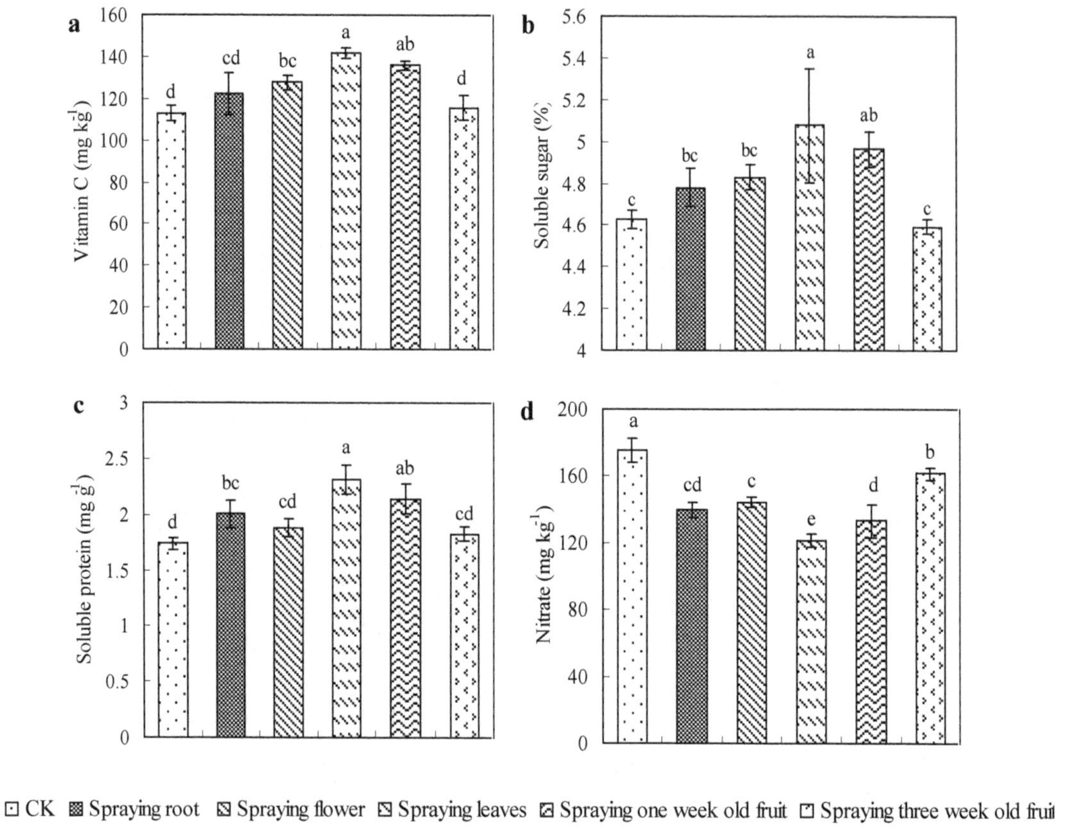

Figure 2. Effects of EM-calcium application on some routine indexes of tomato quality. Columns with the same letter represent values that are not significantly different at the 0.05 level of probability according to the Duncan's multiple range test. Each value is the mean ± SD (n=3). The treatment symbols are the same as the experiment design.

2.2.4. Fruit morphology, yield, and BER incidence

Fruit appearance is the first quality trait to consumers and determined by fruit size, shape and color (Labate, 2007). The tomato fruit size, yield and BER morbidity harvested from different treatments were displayed in Table 4. Results show that different treatments had little effect on the fruit size, maximum D-value of L-diameter, S-diameter, individual fruit weight was 0.46cm, 0.29cm and 17.8g/fruit respectively, which was found between SO and CK, this implied that the fruit size was mainly determined by the genetic cultivar but has few matters with exogenous Ca application. The highest tomato yield of 72.28 t ha⁻¹ was found in SL, followed by SO, CK obtained the lowest yield of 50.26 t ha⁻¹, which was 22.02 t ha⁻¹ lower compared to SL. Overall, SL processed obvious advantages in improving the tomato yield among the treatments. After applied with EM-Calcium, the BER incidence of tomato fruits was well controlled to different extents, according to the results. SL obtained the lowest BER incidence of 9.42% compared to the other treatments, followed by SR, recording as 10.28%. While ST had relatively less influences on the controlling of BER incidence, only 2.23% lower than CK.

Treatment	Fruit size			Yield (t ha⁻¹)	BER incidence (%)
	L-diameter (cm)	S-diameter (cm)	Weight (g/fruit)		
CK	6.18	4.66	116.70	50.26	22.81%
SR	6.42	4.90	128.60	56.42	10.28%
SF	6.43	4.91	128.76	60.84	18.51%
SL	6.60	4.94	133.85	72.28	9.42%
SO	6.64	4.95	134.50	68.09	13.22%
ST	6.40	4.86	126.11	53.38	20.58%

Table 3. Effects of EM-Calcium solution application on tomato fruit size, yield and BER incidence

2.2.5. Ca accumulation in main parts of tomato plant

Ca accumulations in main parts of tomato plant with different EM-Calcium treatments were displayed in Fig. 3; Ca content in upper leaf of SO, ST was significantly higher ($P<0.05$) than that of the other treatments, however, which of SL was slightly lower compared with CK, this indicated that spraying EM-Calcium on tomato leaves may change the migration path of calcium in the leaves; Similar laws were also found in the Ca accumulation of lower leaf, Ca content in lower leaf of SL was significantly lower ($P<0.05$) than that of the other treatments (except CK). According to the results from Ca content in upper leaf and lower leaf, it could be also inferred that SO had the most significant impact on the Ca accumulation of tomato leaves.

ST increased the Ca accumulation of root most significantly, and no significant differences ($P>0.05$) were observed among SF, SL and SO. Ca contents of stem were obviously lower than that of other plant organs, and the differences of which among the treatments were relatively less, Ca content in the stem of SO was significantly higher ($P<0.05$), and there were no significant differences ($P>0.05$) among the other treatments.

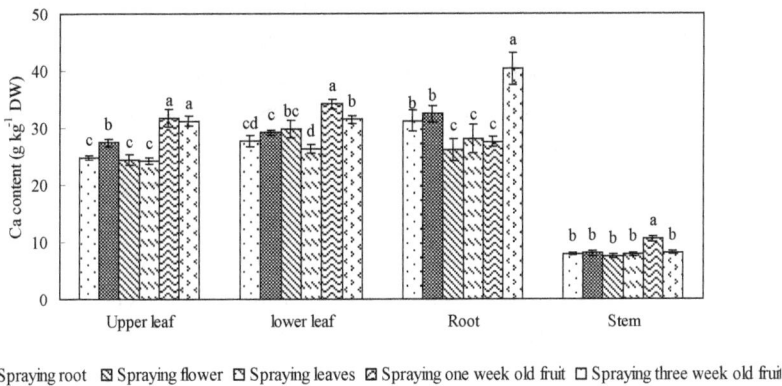

Figure 3. Effects of EM-calcium application on the Ca accumulation in main parts of tomato plants. Columns with the same letter represent values that are not significantly different at the 0.05 level of probability according to the Duncan's multiple range test. Each value is the mean ± SD (n=3). The treatment symbols are the same as the experiment design.

2.3. Discussions

Calcium is well known for having regulatory roles in metabolism and sodium ions may compete with calcium ions for membranebinding sites (Tuna et al., 2007), and is also known to bind to phospholipids and proteins on the membrane surface, which is required to maintain proper membrane structure and integrity (Jones and Lunt, 1967; Suzuki et al., 2003), thus to some degree, high calcium levels can protect the cell membrane from the adverse effects. In view of the important role calcium played, much emphasis was put on the exogenous Ca application for plants. In the choice of Ca nutrient solution for tomato, materials such as $CaSO_4$ $2H_2O$ (Hall, 1977), $CaNO_3$ $4H_2O$ (Eraslan et al., 2007; Murillo-Amador et al., 2006), $CaCl_2$ (Dong Cai-xia, 2001; Schmitz-Eiberger et al., 2002; Xu et al., 2010), CaO (Almeida et al., 2009; Asiegbu and Uzo, 1983) had been selected as Ca resources for different studies. Before preparing the Ca solution in this study, the activity of Ca ions and cost of the calcium solution were mainly taken into consideration during the selection process of materials, we tried to mixed the gypsum with EM previously, while it was found that the solubility of gypsum in EM was not satisfactory compared to that of lime, the lime was finally selected as the Ca resource. Before the experiment we also concerned about whether the Ca^{2+}suspension made by lime would change the survival environment of the acid-loving effective microbes, results proved later that the pH of the mixed liquor fallen back to suitable levels after several days' fermentation. Under microbial actions, the solubility of calcium in EM obtained a satisfactory result (maximum Ca^{2+}solubility of 89.50%), while with the increasing application of lime, EM, and molasses, the dissolving capacity of calcium presented a decline trend, it was predicted that the measured Ca^{2+}concentration of the EM-Calcium solution would stay in a certain value and the solubility of calcium would decrease with the increasing of raw materials. According to the results of the lab experiment, the mixture of Ca^{2+}suspension/ EM/ molasses/ DI water with a volume ratio of 3: 1.5: 1.5: 4 was recommended to practice.

Recent studies tented to adopt foliar-sprayed method on the exogenous Ca application for tomato plants (Eraslan et al., 2007; Gezerel, 1986; Murillo-Amador et al., 2006); there were also experiments (Tabatabaie et al., 2004) about the use of solutions of different concentrations applied to different parts of tomato root system. This experiment showed that the foliar-sprayed method was most beneficial for the Ca accumulation of tomato fruits according to Table 4, spraying EM-Calcium solution on plant leaves increased the Alc-Ca, H_2O-Ca, NaCl-Ca, HAC-Ca contents of tomato fruits more significantly than spraying on the other organs, increment of NaCl-Ca content (calcium pectate) was especially notable, this might be related to the increases of soluble pectin in tomato fruits with the ripening of tomatoes (Ashraf M, 1981). Spraying the calcium nutrition on surface of fruits such as litchi, sweet cherry and grape proved not to be an effective way (Combrink, 1995; Huang et al., 2008; Koffmann, 1996), while for tomato fruits in this experiment, results showed that EM-Calcium application had no significant effects on the Ca increment in old tomato fruit, while which had significant effects on the Ca increment in young tomato fruit.

Vitamin C acts as an antioxidant in plants and its levels are responsive to a variety of environmental or stress factors, for example light, temperature, salt and drought, atmospheric pollutants, metals or herbicides (Singh et al., 2012), and which was reported having positive correlation with potassium (K) supply (Marin, 2009), Bangerth (1976) observed an increase in vitamin C content of tomato fruits treated with calcium chloride. In this experiment, we do not exclude the Ca factor when explained the improvement of tomato quality, while EM was more likely to be the main factor, since EM had been reported to have effects on enhancing crop photosynthesis, increasing crop protein contents, and improving crop quality (Daming, 1999; Shousong, 1998). Another evidence to support this speculation was that the soluble sugar correlated negatively with calcium (Beckles, 2012). Taken as a whole, EM-Calcium application significantly improved the fruit quality of tomato (evaluation indexes including vitamin C, soluble sugar, soluble protein, and nitrate); meanwhile, foliar-spray proved to be a preferable method when supplying EM-Calcium in this study.

The BER incidence may induced by the stresses in the root zone, such as salinity, soil water stress, NH_4^+toxicity and oxygen withholding (Saure, 2001; Tachibana, 1991), although the impact mechanisms are not fully understood, these factors were considered either directly or indirectly related to Ca^{2+}deficiency: Tuna et al.(2007) reported that the exogenous Ca^{2+}application significantly improved growth and physiological variables affected by salt stress, from another perspective, Adams (1990) showed that increasing the salinity above 4mS cm^{-1} by addition of major nutrients would reduce Ca content; According to some studies (Albahou, 1999; Žanić et al., 2011), the increased proportions of NH_4^+in standard nutrient solution were often associated with severity of blossom-end rot known as a physiological disorder of tomato fruit, and Siddiqi (2002) suggested that the NH_4^+presence with a percentage of 10% in total N reduced Ca^{2+}accumulation; The BER occured commonly when the soil moisture content was deficit or fully adequate (Adams, 1992, 1993), and there was a minimum rate of transpiration relative to leaf growth rate below which calcium deficiency symptoms were occured (Hamer, 2003); Tachibana (1991) also reported that withholding the oxygen supply to roots at night was a cause of tomato BER, which greatly inhibited the absorption of Ca. The negative correlation between Ca nutrient supply and BER incidence was also found in this study, similar to many other studies (Besford, 1978; Mestre et al., 2012; Olle M, 2009). It was concluded here that the BER incidence of different EM-Calcium treatments was 2.23%-13.39% lower than that of CK.

Fig. 5 showed the yield of marketable tomatoes, which was significantly increased by the treatments except ST, Gezerel (1986) reported the application of foliar fertilizer containing calcium of 0.2% increased the tomato yields and fruit weight. Mayer (2010) inferred that the effects of EM preparation on crop yield increasing could be related to nutrient inputs by EM carrier substrates. Hu and Qi (2013b) reported that long-term effective microorganisms application could promote crop growth and increase yields and nutrition. In this study, we deduced that EM alone had no obvious effects on crop yield, while it could enhance the availability of nutrient supplied. Mestre et al.(2012) observed a negative and significant correlation between fruit yield and BER (r=-0.810), our results showed that the BER incidence had significant correlation with total tomato yield (r=-0.736) and marketable tomato yield (r=-0.862).

Early study (del Amor and Marcelis, 2006) reported that Ca concentration of tomato plant was significantly reduced by low-Ca supply (0.5 meq L^{-1}) compared with the nutrient standard solution (9 meq L^{-1}), and with 14 days' low-Ca application, Ca concentration in all plant organs (leaves, stems and roots) was reduced by approximately 70% compared to control plants; we reported that 2.0‰ Ca application increased the Ca accumulation in upper leaf, lower leaf, root, stem to maximum rates of 28.09%, 23.50%, 29.15%, 33.34% compared to no-calcium treatment, the causes of the difference were probably related to the nutrient supply methods, calcium supply through roots increased the Ca content of tomato fruits by increasing which of other organs, while calcium spray on leaves or young fruits increased Ca content of tomato fruits by changing the Ca migration, the evidence to support this speculation in this study was that the foliar spray increased the Ca content of fruits but decreased which of leaves. Dong (2001) guessed that a "Ca-attracted" center was formed in the spray organ when spraying Ca nutrient ($CaCl_2$), Ca^{2+} was attracted to the center and then migrated from the center to the organ which needed Ca most. However, the migration mechanism about Ca migration with exogenous Ca application needed to be examined by [45]Ca tracing technique (Behling et al., 1989; Yamauchi et al., 1986).

3. Effects of EM-Calcium on production of flue-cured tobacco

3.1. Materials and methods

3.1.1. Test site

The experiments were carried out in a plastic sheet covered greenhouse from March 2013 to September 2013 in the Vegetables and Flowers Institute of Nanjing (latitude 31°43' N, longitude 118°46'E), China. The average annual rainfall is about 1106.5mm, with the rainy season from the end of June to the middle of July, and the average yearly temperature is approximately 15.7°C and average humidity is about 81%. The soil characters were: pH 5.68, 14.47 g/kg organic matter, 28.5 mg/kg available P, 153.84 mg/kg available K, 1.3 8g/kg total nitrogen.

3.1.2. Experimental design

K326 was chosen as the flue-cured tobacco plant material, with young plants elaborately cultivated in made-in-order seedling trays; then they would be transplanted into the lysimeters when they grew 6 expanded leaves. The planting density was 12 plants per treatment, with the line spacing of 0.8m and plant spacing of 0.6m. After that, conventional field management was conducted in the first week.

For simulating the water stress in growing stages, the irrigation amount was designed as 400mm during the whole growth stage. Irrigation waters of root-extending stage, vigorous stage and maturity stage in this experiment were assigned as 40%, 20%, 40% for the total water consumption respectively, and they were irrigated 6 times in each growth stage. Tobacco dedicated fertilizers (provided by the Institute of Guizhou Tobacco Science, N: P_2O_5: K_2O=1:2:3) were applied according to a proportion of basal dressing: topdressing=7:3, the latency time of topdressing was 26 days after transplanted.

Detailed experimental treatments regarding the water-retaining agent amount and EM-calcium content were shown in Table 4. EM-calcium with 2.0‰ Ca^{2+}concentration was prepared as 2.2.1.

Treatment	Ca²⁺ concentration	EM-Ca amount (ml/time)	Intervals of spraying (days)	MP3005 (g/plant)	Growth stage
CK	---	---	---	---	---
P1	1‰	2	3	0	Root extending (R)
P2	1‰	2	3	0	Vigorous (V)
P3	1‰	2	3	0	R+V
P4	1‰	2	3	30	R
P5	1‰	2	3	30	V
P6	1‰	2	3	30	R+V

Table 4. Experimental design

3.1.3. Measurements

3 representative tobacco plants were sampled for each replication. At harvest time, the lower leaves, middle leaves and upper leaves were collected orderly, killed by 105℃ high temperature and toasted to the constant weight (Hou et al., 2012; Maomao Hou, 2013). Weight of dry tobacco leaves was measured and recorded to calculate the total yield.

IWUE was calculated by the formula (Aujla et al., 2005; Ünlü et al., 2011):

$$IWUE = \frac{Y}{I}$$

Where, IWUE (kg/m^3) was the irrigation water use efficiency; Y was the tobacco dry yield (kg/hm^2); I was the irrigation amount (m^3/hm^2).

3.2. Results

3.2.1. Agronomic characters of flue-cured tobacco

Fig. 4 showed the changes of total area of flue-cured tobacco leaves in single plant with days after transplanted. As shown in the Figure 4, during 45~77 days, flue-cured tobaccos in P4 grew more satisfactory compared to that in other treatments, and the leaf area of single plant in T4 reached a higher value of 36897.9 cm² in 77 days. However, P2 showed a poor performance in leaf area enhancement, recording as 32110.1 cm² only. From the view of leaf area enhancement, the effects of water-retaining agent treatments were obviously better than those with no water-retaining agent. Under the same application amount of water-retaining agent, the effects of EM-calcium spray in root-extending stage were better than those in vigorous stage and root-extending+vigorous stage. The results were different from the expectation, it was maybe that the EM-calcium spray blocked the leaf stoma after long-time application. On the whole, P4 was supposed to be the better treatment from a pure view of leaf area increasing.

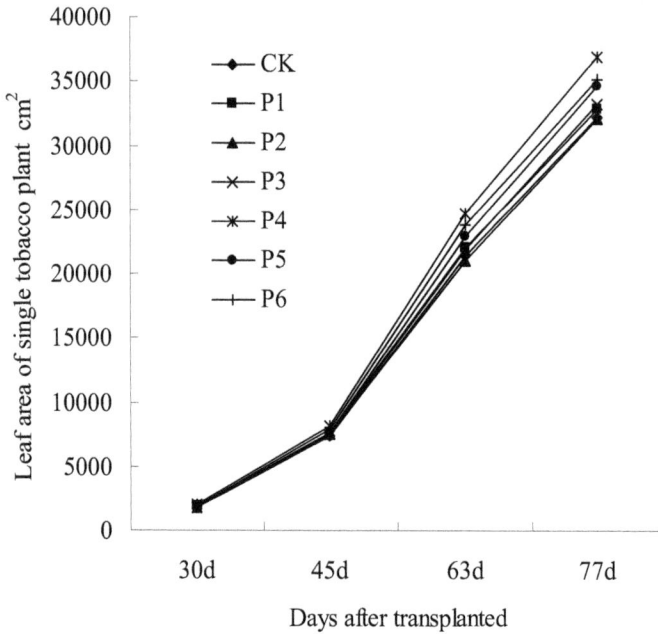

Figure 4. Leaf area increasing with days after transplanted

Table 5 displayed the plant height and stem girth with days after transplanted under different treatments. As shown from the Table 5, no obvious differences were found among the treatments, this was mainly because the fertilizer application was equal in different treatments. At 77 days after transplanted, the plant height of different treatments were recorded as 105.4~122.0 cm and the plant stem girth were around 6.9~7.4 cm.

Treatment	Height (cm)				Stem (cm)			
	30d	45d	63d	77d	30d	45d	63d	77d
CK	9.2	38.7	84.5	109.9	3.1	4.3	6.0	7.0
P1	10.5	39.6	86.2	110.6	3.2	4.3	6.3	7.2
P2	10.2	35.8	83.9	105.4	3.1	4.0	5.9	6.9
P3	11.1	41.2	88.0	113.9	3.3	4.4	6.2	7.3
P4	12.8	43.9	97.4	122.0	3.7	4.8	6.5	7.4
P5	11.4	41.2	91.8	108.5	3.4	4.5	6.3	7.1
P6	12.3	42.5	96.9	119.8	3.4	4.6	6.6	7.2

Table 5. Plant stem girth and height with days after transplanted

3.2.2. Chlorophyll content of tobacco leaves

Fig. 5 showed the dynamic changes of chlorophyll content in tobacco leaves under different treatments more visually, chlorophyll content in tobacco leaves was decreased with a higher rate at the later growth stage. The chlorophyll content of tobacco leaves with water-retaining agent was slightly lower than that with no water-retaining agent, at 63 d after transplanted, the chlorophyll content of tobacco leaves under different treatments was recorded as 1.69 mg/g-1.85 mg/g; at 87 days after transplanted, the chlorophyll content of tobacco leaves was in the lowest level of 0.97 mg/g-1.12 mg/g.

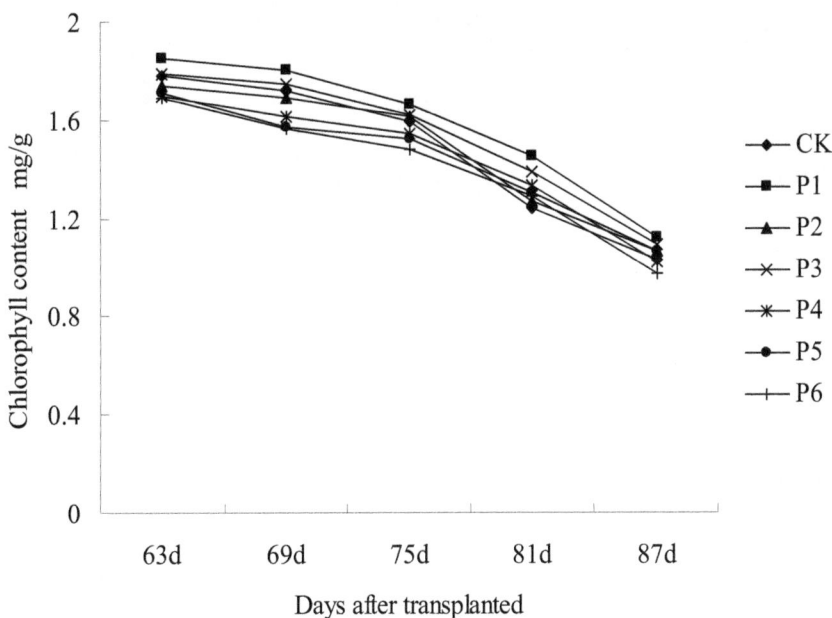

Figure 5. The changes of chlorophyll content in tobacco leaves varying with days after transplanted

3.2.3. Dry matter accumulation

Table 6 showed the dry matter accumulation of different tobacco organs and the total yield under different treatments. From the table it was found that the dry matter amount of leaf and stem and root in P4 was the highest, recording as 73.6 g, 88.1 g and 164.9 g respectively. The dry matter amount of leaves with water-retaining agent treatments were 4.99%-14.28% higher than that with no water-retaining agent application. Under the same amount of water--retaining agent application, EM-calcium spray in root-extending stage was much better than that in other stages. Dry matter amount of leaf in P1 and P4 was higher than that of other treatment, and the tobacco yield of P4 was highest, recording as 2473.5 kg/hm², followed closely by P6, which of P2 was the lowest.

Treatment	Root		Stem		Leaf		Whole Plant(g)	Yield (kg/hm²)
	Dry matter(g)	Proportion (%)	Dry matter (g)	Proportion (%)	Dry matter (g)	Proportion(%)		
CK	63.2	21.76%	82.9	28.55%	144.3	49.69%	290.4	2164.5
P1	62.1	21.01%	85.4	28.89%	148.1	50.10%	295.6	2221.5
P2	58.7	21.43%	75.4	27.53%	139.8	51.04%	273.9	2097.0
P3	64.5	22.13%	80.5	27.62%	146.5	50.26%	291.5	2197.5
P4	73.6	22.54%	88.1	26.97%	164.9	50.49%	326.6	2473.5
P5	67.7	22.76%	78.3	26.32%	151.5	50.92%	297.5	2272.5
P6	68.7	22.50%	77.9	25.52%	158.7	51.98%	305.3	2380.5

Table 6. The dry matter accumulation of different tobacco organs and the total yield under different treatments.

3.2.4. Nutrient absorption

Calcium is well known to maintain proper membrane structure and integrity and plays important roles in crop development and disease control (Almeida et al., 2009; Berry et al., 1988; Evans, 1953; Hall, 1977). In flue-cured tobacco cultivation, calcium helps to coordinate the physiology function of tobacco plant, making the tobacco plant root system stronger, growing vigorously and harvesting timely. Ca^{2+} has promoting effects for the growth of tobacco seedlings and can improve the drought-resistant ability of tobacco seedlings. In addition, the high calcium content in tobacco leaves delays the maturity of tobacco, characterized by stiffness and hardness of the tobacco leaves, thus the use value of tobacco leaves is decreased, excess calcium may also cause disorder of some microelements in tobacco plants and produce toxic impacts; while calcium deficiency can lead to the deformity of tobacco plants and generate a spoon-shaped reverse disease of tobacco leaves. As was shown in Fig. 6, the exogenous application of calcium significantly increased the calcium content in tobacco leaves, and the effects of spraying in root-extending stage were more satisfactory than those in other stages, this may be related to that the water stress in maturity affected negatively the effects of EM-calcium.

3.2.5. Irrigation water use efficiency

Fig. 7 showed the *IWUE* of flue-cured tobacco plants with different treatments. It could be seen that the differences of *IWUE* among the treatments were more significant, P4 obtained the highest *IWUE* of 0.618 kg/m³, followed by P6, and *IWUE* value of T2 was the minimum, recording as 0.524 kg/m³. Since the equal irrigation amount among the treatments, *IWUE* presented a positive relationship with the flue-cured tobacco yield.

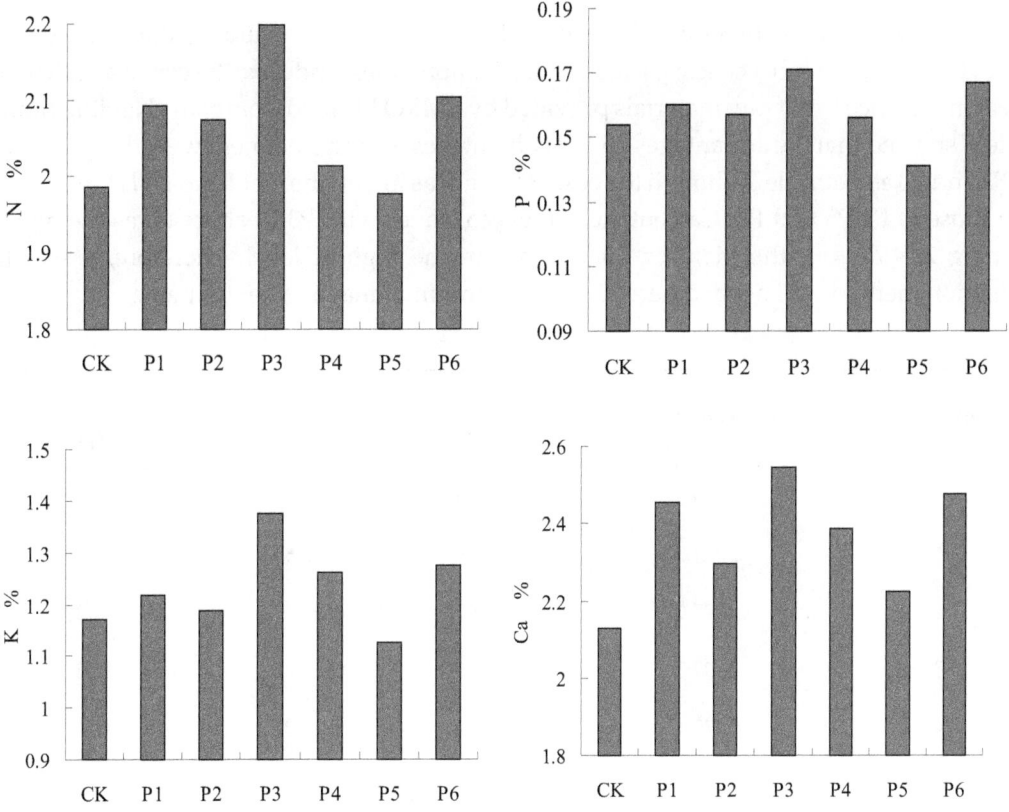

Figure 6. Nutrient absorption of tobacco leaves with different treatments

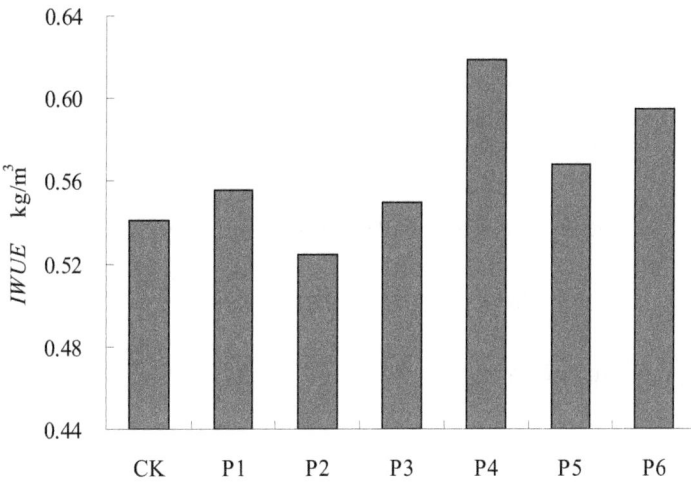

Figure 7. Irrigation water use efficiency with different treatments

3.2.6. Optimal selection of the best management scheme

Table. 7 showed the evaluation indexes of different treatments including the tobacco yield, *IWUE*, Ca content in tobacco leaves and the cost, among these indexes, the cost was calculated based on the price list of raw materials provided by EMRO Limited Company, Nanjing Branch. Table. 7 showed that the advantages and disadvantages of each treatment were distinct, taking P1, P2 and P3 as example, although the cost of them was lower, the yield and *IWUE* were lower than those of P4,P5 and P6; Ca content of tobacco leaves with P3 treatment was the highest, reaching 2.546%, but the yield of P3 was not in the highest level. Therefore, a scientific evaluation method was needed here to select the optimal management scheme.

Treatment	Yield (kg/hm²)	IWUE (kg/m³)	Ca (%)	Cost (USD/plant)
CK	2164.5	0.54	2.131	0
P1	2221.5	0.56	2.454	0.0033
P2	2097.0	0.52	2.297	0.0033
P3	2197.5	0.55	2.546	0.0067
P4	2473.5	0.62	2.387	0.0167
P5	2272.5	0.57	2.224	0.0167
P6	2380.5	0.60	2.478	0.0200

Table 7. Evaluation indexes

Modeling approach was shown below (Chen and Li, 2010; Chou et al., 2012):

Supposing that there were n evaluation indexes and m schemes, m schemes corresponding with n indexes obtained the following matrix:

$$R = (r_{ij})_{m \times n}$$

Where; r_{ij} is the jth evaluation index of the ith scheme. To r_j, there was information entropy:

$$E_j = -\sum_{i=1}^{m} p_{ij} \ln p_{ij}, \ (j=1, 2, 3, \ldots n)$$

And P_{ij} were calculated as the formula:

$$p_{ij} = r_{ij} \left/ \sum_{i-1}^{m} r_{ij} \right.$$

The entropy value of jth index was:

$$e_j = \frac{1}{\ln m} E_j, \ (j=1, 2, 3, \ldots, n)$$

The objective weight of jth index was:

$$\theta_j = (1-e_j) \Big/ \sum_{i=1}^{n} (1-e_j), \quad (j=1, 2, 3, \ldots, n)$$

It was clear that:

$$0 \le \theta_j \le 1; \quad \sum_{j=1}^{n} \theta_j = 1$$

This study took the subjective information into the calculations, the comprehensive weight could be obtained by combining the subjective weight $w_1, w_2, w_3 \ldots w_n$ of the decision makers with the objective weight θ_j ($j=1,2,3,\ldots,n$):

$$\alpha_j = \theta_j \overline{\omega}_j \Big/ \sum_{j=1}^{n} \theta_j \overline{\omega}_j, \quad (j=1, 2, 3, \ldots, n)$$

Recording the optimum value of each row as r_j^*, normalize the elements in the matrix, and r_j^* value was varied with the index characters. The indexes could be divided into two classes: The profitable indexes and the damnous indexes, which were "the larger the better" and "the smaller the better", listing as follows:

$$d_{ij} = \begin{cases} \dfrac{r_{ij}}{r_j^*}, & r_j^* = \max\{r_{ij}\} \\[2mm] \dfrac{r_j^*}{r_{ij}}, & r_j^* = \min\{r_{ij}\} \end{cases}$$

The entropy coefficient value (A better management scheme would obtain a higher entropy coefficient value) of each treatment could be calculated by:

$$\lambda_i = \sum_{j=1}^{n} \alpha d_{ij}, \quad i=1, 2, 3, \ldots, m.$$

The calculated objective weight of the tobacco yield, $IWUE$, Ca content in tobacco leaves and the cost were 0.2142, 0.2135, 0.2003, and 0.3720. However, since the strict requirement on water saving, the subjective weight of which was assigned as 0.4, 0.4, 0.15, 0.05 respectively. Fig. 8 showed the entropy weight coefficient of different treatments, based on the principle of entropy weight coefficient "the higher the better", P4 was supposed to be the best scheme, in other words, 30 g/plant MP3005 water-retaining agent combined with EM-calcium spray (‰1 Ca^{2+}) during the root-extending stage of flue-cured tobaccos was the optimal management scheme, and the interval of spraying time was 3 days with 2 mm each time on the back side of tobacco leaves. Additionally, entropy weight coefficient value of P1 and P4 were similar, but the mechanism was different, P1 tended to obtain a lower cost, and P4 tended to obtain a higher yield and $IWUE$.

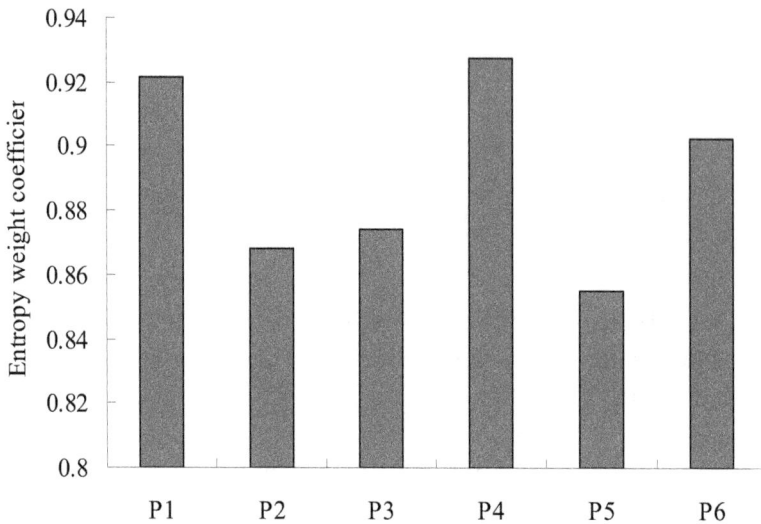

Figure 8. Entropy weight coefficient of different treatments

3.3. Conclusions

Treatment P4 (spraying 1‰ EM-Ca amount 2 ml with 30 g MP3005 per plant once in 3 days during tobacco root extending stage) obtained the highest flue-cured tobacco yield of 2473.5 kg/hm², followed by P6, and yield of P2 was lowest, recording as only 2097.0 kg/hm². The irrigation water use efficiency of P4 was highest, reaching 0.618 kg/m³. Exogenous Ca supply significantly increased the Ca content in tobacco leaves. The evaluation results of entropy weight coefficient evaluation model showed that P4 was the best management scheme, that was to say, 30 g/plant MP3005 water-retaining agent combined with EM-calcium spray (1‰ Ca²⁺) during the root-extending stage of flue-cured tobaccos was the optimal management scheme, and the interval of spraying time was 3 days with 2ml each time on the back side of tobacco leaves.

4. General conclusions

The research results showed that **Active** EM-Calcium could promote crop growth, improve the yield and disease resistance of crops. Main conclusions could be drawn as below:

1. The application of **Active** EM-Calcium increased the Ca accumulation in upper leaf, lower leaf, root and stem to maximum rates of 28.09%, 23.50%, 29.15%, 33.34% compared to no-calcium treatment. The BER incidence of **Active** EM-Calcium treatments was lower than that of CK. The BER incidence had significantly negative correlation with total tomato yield (r=-0.736) and marketable tomato yield (with no BER) (r=-0.862).

2. Exogenous Ca supply significantly increased the Ca content in tobacco leaves. Treatment P4 (spraying 1‰ EM-Ca amount 2 ml with 30 g MP3005 per plant once in 3 days during tobacco root extending stage) obtained the highest flue-cured tobacco yield of 2473.5 kg/hm^2, and the highest irrigation water use efficiency with 0.618 kg/m^3.

Acknowledgements

This work was financed by the Fund of the Ministry of Water Resources Public Welfare Project (201301017), the Fundamental Research Funds for the Central Universities (2013/B13020057), the Graduate Student Research and Innovation Program of Jiangsu Province (257) and Creative Funds of Jiangsu province (CXZZ13_0266).

Author details

Xiaohou Shao[1,2], Tingting Chang[1,3], Maomao Hou[1,3], Yalu Shao[3] and Jingnan Chen[3]

1 Key Laboratory of Efficient Irrigation-Drainage and Agricultural Soil-Water Environment in Southern China of Ministry of Education, Hohai University, Nanjing, PR China

2 Nanjing Ning-ya Environmental Science and Technology Limited Company, Nanjing, PR China

3 College of Water Conservancy and Hydropower Engineering, Hohai University, Nanjing, PR China

References

[1] Adams, P., 1990. Effect of salinity on the distribution of calcium in tomato (Lycopersicon esculentum) fruit and leaves, In: Beusichem, M.L. (Ed.), Plant Nutrition — Physiology and Applications. Springer Netherlands, . 473-476.

[2] Adams, P., Ho, L.C., 1992. The susceptibility of modern tomato cultivars to blossom-end rot in relation to salinity. J. Hor. Sci. 67, 827-839.

[3] Adams, P., Ho, L.C., 1993. Effects of environment on the uptake and distribution of calcium in tomato and on the incidence of blossom-end rot. Plant Soil. 154, 127-132.

[4] Albahou, M., 1999. Alternative greenhouse tomato production system. ph. D. Thesis. Oregon State University. Oregon, 223 .

[5] Almeida, G.D., Pratissoli, D., Zanuncio, J.C., Vicentini, V.B., Holtz, A.M., Serrão, J.E., 2009. Calcium silicate and organic mineral fertilizer increase the resistance of tomato plants to Frankliniella schultzei. Phytoparasitica. 37, 225-230.

[6] AOAC, 1990. Official Methods of Analysis, 15th ed. Association of Official Analytical Chemists. Washington, DC.

[7] Ashraf, M. K.N., 1981. Studies on the pectinesterase activity and some chemical constituents of some Pakistan mango varieties during storage ripening. J. Agric. Food Chem. 29, 526-528.

[8] Asiegbu, J.E., Uzo, J.O., 1983. Effects of lime and magnesium on tomato (Lycopersicon esculentum Mill) grown in a ferrallitic sandy loam tropical soil. Plant Soil. 74, 53-60.

[9] Aujla, M.S., Thind, H.S., Buttar, G.S., (2005). Cotton yield and water use efficiency at various levels of water and N through drip irrigation under two methods of planting. Agricultural Water Management. 71, 167-179.

[10] Bangerth, F., 1976. Relationship between calcium content and the content of ascorbic acid in apple, pear and tomato fruits. Qual. Plant. 26, 341-348.

[11] Besford, R.T., 1978. Effect of potassium nutrition of three tomato varieties on incidence of blossom-end rot. Plant Soil. 50, 179-191.

[12] Berry, S.Z., Madumadu, G.G., Uddin, M.R., 1988. Effect of calcium and nitrogen nutrition on bacterial canker disease of tomato. Plant Soil. 112, 113-120.

[13] Behling, J., Gabelman, W.H., Gerloff, G.C., 1989. The distribution and utilization of calcium by two tomato (Lycopersicon esculentum Mill.) lines differing in calcium efficiency when grown under low-Ca stress. Plant Soil. 113, 189-196.

[14] Bennett, A.J., Mead, A., Whipps, J.M., 2009. Performance of carrot and onion seed primed with beneficial microorganisms in glasshouse and field trials. Biol Control. 51, 417-426.

[15] Beckles, D.M., 2012. Factors affecting the postharvest soluble solids and sugar content of tomato (Solanum lycopersicum L.) fruit. Postharvest Biol Tec. 63, 129-140.

[16] Berry, S.Z., Madumadu, G.G., Uddin, M.R., (1988). Effect of calcium and nitrogen nutrition on bacterial canker disease of tomato. Plant Soil. 112, 113-120.

[17] Chen, T.Y., Li, C.H., 2010. Determining objective weights with intuitionistic fuzzy entropy measures: A comparative analysis. Information Sciences. 180, 4207-4222.

[18] Chou, Y.C., Yen, H.Y., Sun, C.C., 2012. An integrate method for performance of women in science and technology based on entropy measure for objective weighting. Qual Quant, 1-16.

[19] Chiasson, D., Ekengren, S., Martin, G., Dobney, S., Snedden, W., 2005. Calmodulin-like Proteins from Arabidopsis and Tomato are Involved in Host Defense Against Pseudomonas syringae pv. tomato. Plant Mol Biol. 58, 887-897.

[20] Chung, M., Han, J.S., Giovannoni, J., Liu, Y., Kim, C., Lim, K., Chung, J., 2010. Modest calcium increase in tomatoes expressing a variant of Arabidopsis cation/H+ antiporter. Plant Biotechnol Rep. 4, 15-21.

[21] Combrink, N.J.J., Jacobs, G., Maree, P.C.J., 1995. The effect of calcium and boron on the quality of muskmelons (Cucumis melo L.). Journal of Southern African Society for Horticultural Science. 5, 33-38.

[22] Daly, M.J., Stewart, D.P.C., 1999. Influence of "Effective Microorganisms" (EM) on Vegetable Production and Carbon Mineralization–A Preliminary Investigation. J. Sustain Agr. 14, 15-25.

[23] Daming, R., 1999. Studies on the action mechanism of effective microorganisms in crop production. Journal of maize science in China. 7, 62-64.

[24] Del Amor, F.M., Marcelis, L.F.M., 2006. Differential effect of transpiration and Ca supply on growth and Ca concentration of tomato plants. Sci Hortic. 111, 17-23.

[25] Dong C.X, Z.J.M., Fan X.H., Wang H.Y. 2001. Effects of different ways of Ca supplements on the Ca content and forms in mature fruits of tomato. Plant Nutrition and Fertilizer Science. 10, 91-95.

[26] Dorais, M., Papadopoulos, A., Gosselin, A., 2001. Greenhouse tomato fruit quality. Hort Reviews. 26, 239-319.

[27] Eraslan, F., Akbas, B., Inal, A., Tarakcioglu, C., 2007. Effects of foliar sprayed calcium sources onTomato mosaic virus (ToMV) infection in tomato plants grown in greenhouses. Phytoparasitica. 35, 150-158.

[28] Evans, H.J., 1953. Relation of calcium nutrition to the incidence of blossom-end rot in tomatoes. proc. Am. Soc. Hor. Sci. 61, 346-352.

[29] Freitas, S., Jiang, C.Z., Mitcham, E., 2012. Mechanisms Involved in Calcium Deficiency Development in Tomato Fruit in Response to Gibberellins. J Plant Growth Regul. 31, 221-234.

[30] Gezerel, Ö., 1986. The Effect of Calcium-Containing Foliar Fertilizers on Tomato Yields, In: Alexander, A. (Ed.), Foliar Fertilization. Springer Netherlands, . 304-309.

[31] Hall, D.A., 1977. Some effects of varied calcium nutrition on the growth and composition of tomato plants. Plant Soil. 48, 199-211.

[32] Hamer, P.J.C., 2003. Analysis of strategies for reducing calcium deficiencies in glasshouse grown tomatoes: model functions and simulations. Agr Syst. 76, 181-205.

[33] Heo, S.-U., Moon, S.-Y., Yoon, K.-s., Kim, Y.-J., Koo, Y.-M., 2008. Enhanced compost maturity by effective microorganisms. J Biotechnol. 136, Supplement, S65.

[34] Higa, T., 1997. Great revolution to save the earth in Chinese edition. China Agriculture Publisher University.

[35] Hou, M.M., Shao, X.H., Chen, L.H., Chang, T.T., Wang, W.N., Yang, Q., Wang, Y.F., 2012. Study on nitrogen utilization efficiency of flue-cured tobacco with 15n tracing technique, 1st International Conference on Energy and Environmental Protection, 520-524.

[36] Hou, M.M., Shao X.H., Zhai Y.M. 2013. Entropy weight coefficient evaluation of comprehensive index for flue-cured tobacco and its response to different water-nitrogen treatments. Research on Crops. 14, 1232-1237.

[37] Huang, X.M., Wang, H.C., Zhong, W.L., Yuan, W.-Q., Lu, J.M., Li, J.G., 2008. Spraying calcium is not an effective way to increase structural calcium in litchi pericarp. Sci Hortic. 117, 39-44.

[38] Hu, C., Qi, Y., 2013a. Effective microorganisms and compost favor nematodes in wheat crops. Agron. Sustain. Dev., 1-7.

[39] Hu, C., Qi, Y., 2013b. Long-term effective microorganisms application promote growth and increase yields and nutrition of wheat in China. European Journal of Agronomy. 46, 63-67.

[40] Jones, R.G.W., Lunt, O.R., 1967. The function of calcium in plants. Bot. Rev. 33, 407-426.

[41] Javaid, A., 2010. Beneficial Microorganisms for Sustainable Agriculture, In: Lichtfouse, E. (Ed.), Genetic Engineering, Biofertilisation, Soil Quality and Organic Farming. Springer Netherlands, pp. 347-369.

[42] Khaliq, A., Abbasi, M.K., Hussain, T., 2006. Effects of integrated use of organic and inorganic nutrient sources with effective microorganisms (EM) on seed cotton yield in Pakistan. Bioresource Technology. 97, 967-972.

[43] Koffmann, W., Wade, N.L., Nicol, H., 1996. Tree sprays and root pruning fail to control rain induced cracking of sweet cherries. Plant Protection Quarterly. 11, 126-130.

[44] Kumar, S., Dey, P., 2011. Effects of different mulches and irrigation methods on root growth, nutrient uptake, water-use efficiency and yield of strawberry. Scientia Horticulturae. 127, 318-324.

[45] Labate, J.A., Grandillo, S., Fulton, T., Mun´ os, S., Caicedo, A., Peralta, I., et al., 2007. Tomato. In: Kole, C. (Ed.),. Genome Mapping and Molecular Breeding in Plants. vol. 5. Springer, New York, pp. , 1–125.

[46] Lima-Silva, V., Rosado, A., Amorim-Silva, V., Muñoz-Mérida, A., Pons, C., Bombarely, A., Trelles, O., Fernández-Muñoz, R., Granell, A., Valpuesta, V., Botella, M., 2012.

Genetic and genome-wide transcriptomic analyses identify co-regulation of oxidative response and hormone transcript abundance with vitamin C content in tomato fruit. BMC Genomics. 13, 1-15.

[47] Marin, A., Rubio, J. S., Martinez, V., & Gil, M. I., 2009. Antioxidant compounds in green and red peppers as affected by irrigation frequency, salinity and nutrient solution composition. J Sci Food Agr. 89, 1352-1359.

[48] Mayer, J., Scheid, S., Widmer, F., Fließbach, A., Oberholzer, H.R., 2010. How effective are 'Effective microorganisms® (EM)'? Results from a field study in temperate climate. Appl Soil Ecol. 46, 230-239.

[49] Mestre, T.C., Garcia-Sanchez, F., Rubio, F., Martinez, V., Rivero, R.M., 2012. Glutathione homeostasis as an important and novel factor controlling blossom-end rot development in calcium-deficient tomato fruits. J Plant Physiol. 169, 1719-1727.

[50] Mowa, E., Maass, E., 2012. The effect of sulphuric acid and effective micro-organisms on the seed germination of Harpagophytum procumbens (devil's claw). S Afr J Bot. 83, 193-199.

[51] Murillo-Amador, B., Jones, H.G., Kaya, C., Aguilar, R.L., García-Hernández, J.L., Troyo-Diéguez, E., Ávila-Serrano, N.Y., Rueda-Puente, E., 2006. Effects of foliar application of calcium nitrate on growth and physiological attributes of cowpea (Vigna unguiculata L. Walp.) grown under salt stress. Environ Exp Bot. 58, 188-196.

[52] Ohat Y, Y.K., Deguchi M, 1970. Chemical fractionation of calcium in the fresh leaf blade and influences of deficiency or over supply of calcium and age of leaf on the content of each calcium fraction. J. Sci. Soil Manure. 41, 19-26.

[53] Olle M, B.I., 2009. Causes and control of calcium deficiency disorders in vegetables: a review. J Hort Sci Biotechnol. 84, 577-584.

[54] Raleigh, S.M., Chucka, J.A., 1944. Effect of nutrient ratio and concentration on growth and composition of tomato plants and on the occurrence of blossom-end rot of the fruit. Plant Physiol. 19, 671-678.

[55] Sachan, R.S., Sharma, R.B., 1981. An easy estimation of tomato root parameters based on calcium absorption. Biol Plant. 23, 311-314.

[56] Shao X.H., H.M.M., .Chen .J.N. 2013. Effects of EM-calcium spray on Ca uptake, blossom-end rot incidence and yield of greenhouse tomatoes (Lycopersicon esculentum). Research on Crops. 14, 1159-1166.

[57] Shousong, Y., 1998. The effect of effective microorganisms spray on soybean yield and quality. Liaoning agricultural science in China. 3, 12-13.

[58] Saure, M.C., 2001. Blossom-end rot of tomato (Lycopersicon esculentum Mill.) — a calcium- or a stress-related disorder? Sci Hortic. 90, 193-208.

[59] Schmitz-Eiberger, M., Haefs, R., Noga, G., 2002. Calcium deficiency - Influence on the antioxidative defense system in tomato plants. J Plant Physiol. 159, 733-742.

[60] Siddiqi, M.Y., Malhotra, B., Min, X., Glass, A.D.M, 2002. Effects of ammonium and inorganic carbon enrichment on growth and yield of a hydroponic tomato crop. Plant Nutr. Soil Sci. 165, 191-197.

[61] Suzuki, K., Shono, M., Egawa, Y., 2003. Localization of calcium in the pericarp cells of tomato fruits during the development of blossom-end rot. Protoplasma. 222, 149-156.

[62] Sonneveld, C., Voogt, W., 2009. Calcium Nutrition and Climatic Conditions, Plant Nutrition of Greenhouse Crops. Springer Netherlands, 173-201.

[63] Singh, D.P., Beloy, J., McInerney, J.K., Day, L., 2012. Impact of boron, calcium and genetic factors on vitamin C, carotenoids, phenolic acids, anthocyanins and antioxidant capacity of carrots (Daucus carota). Food Chem. 132, 1161-1170.

[64] Tachibana, S., 1991. Import of calcium by tomato fruit in relation to the day-night periodicity. Sci Hortic. 45, 235-243.

[65] Tabatabaie, S.J., Gregory, P.J., Hadley, P., 2004. Uneven distribution of nutrients in the root zone affects the incidence of blossom end rot and concentration of calcium and potassium in fruits of tomato. Plant Soil. 258, 169-178.

[66] Taylor, M.D., Locascio, S.J., 2004. Blossom-end rot: a calcium deficiency. J. Plant Nutr. 27, 123-139.

[67] Truax, B., Gagnon, D., 1993. Effects of straw and black plastic mulching on the initial growth and nutrition of butternut, white ash and bur oak. Forest Ecology and Management. 57, 17-27.

[68] Tuna, A.L., Kaya, C., Ashraf, M., Altunlu, H., Yokas, I., Yagmur, B., 2007. The effects of calcium sulphate on growth, membrane stability and nutrient uptake of tomato plants grown under salt stress. Environ Exp Bot. 59, 173-178.

[69] Ünlü, M., Kanber, R., Koç, D.L., Tekin, S., Kapur, B., 2011. Effects of deficit irrigation on the yield and yield components of drip irrigated cotton in a mediterranean environment. Agricultural Water Management. 98, 597-605.

[70] Wang, F., Kang, S., Du, T., Li, F., Qiu, R., 2011. Determination of comprehensive quality index for tomato and its response to different irrigation treatments. Agr Water Manage. 98, 1228-1238.

[71] Xu, T., Li, T., Qi, M., 2010. Calcium effects on mediating polygalacturonan activity by mRNA expression and protein accumulation during tomato pedicel explant abscission. Plant Growth Regul. 60, 255-263.

[72] Yamauchi, T., Hara, T., Sonoda, Y., 1986. Effects of boron deficiency and calcium supply on the calcium metabolism in tomato plant. Plant Soil. 93, 223-230.

[73] Yamazaki, H. H., T., 1995. Calcium nutrition affects resistance of tomato seedlings to bacterial wilt. HortScience. 30, 91-93.

[74] Yang, L., Qu, H., Zhang, Y., Li, F., 2012. Effects of partial root-zone irrigation on physiology, fruit yield and quality and water use efficiency of tomato under different calcium levels. Agr Water Manage. 104, 89-94.

[75] Zhou, S., Wei, C., Liao, C., Wu, H., 2008. Damage to DNA of effective microorganisms by heavy metals: Impact on wastewater treatment. J Environ Sci. 20, 1514-1518.

[76] Zhou, Q., Li, K., Jun, X., Bo, L., 2009. Role and functions of beneficial microorganisms in sustainable aquaculture. Bioresource Technol. 100, 3780-3786.

[77] Žanić, K., Dumičić, G., Škaljac, M., Ban, S.G., Urlić, B., 2011. The effects of nitrogen rate and the ratio of NO_3^-:NH_4^+ on Bemisia tabaci populations in hydroponic tomato crops. Crop Prot. 30, 228-233.

4

Organic Agriculture, Sustainability and Consumer Preferences

Terrence Thomas and Cihat Gunden

1. Introduction[1]

Scholars acknowledge that early man provided food for himself and his family via gathering what was available to him in his surroundings; he relied on nature for his sustenance. As hunter gatherers, man lacked the capacity to manipulate the environment to produce food beyond the amount that was available naturally. Consequently, there was minimal or no environmental impact, the human population remained small and in balance with nature; hunter gatherers' population could not expand beyond the available sources of food [1-3]. Over time, however, as hunter gatherers learn to cope with their environment and became more adept at gathering food, the population increased, leading to the next stage in the evolution of the food production system—the Neolithic revolution or the development of agriculture. The development of agriculture led to sedentary communities, increase in population size and the specialization of labor, all of which facilitated technological development, i.e., improved tools, dwellings and means for transporting water and materials. In sum, man learned and applied techniques for domesticating animals and plants, or put another way, agriculture was invented. Yet, at this early stage in the practice of agriculture, man's interaction with his sustenance base could be described as "give and take"; a relationship in which man essentially learned from his experience living in the environment, a sort of 'symbiotic" relationship with his sustenance base that resulted in little or no adverse environmental impact. Even when there was adverse impact, the population was small and technology environmentally benign, which allowed the sustenance base to recover. The invention of agriculture laid the foundation for the development of civilization, increase in knowledge and man's capability to manipulate the environment. It was not until the birth of modern science and its application to the development of

---

1 This section of the chapter is drawn extensively on the work of [4-5].

techniques for producing goods and services that man acquired the capability to manipulate the environment for producing food to meet his needs. The birth of modern science, following the Enlightenment, nurtured a culture that promoted and reinforced the world view that man through the application of science would be able to master and manipulate the environment to meet his needs. Advances in science during this era (17th and 18th century) led to the Industrial Revolution and the progressive industrialization of agriculture.

Prior to the intensive application of science to agriculture, the production of food and fiber relied on what is now referred to as traditional methods, which included: crop rotation, organic manure from animals and cover crops, animal power, intensive use of labor on small farms and a conventional artisan approach to plant and animal improvement—agriculture relied heavily on natural process, i.e., the ecology in which it was nested. Thus, in terms of today's language food production was substantively organic. The industrial revolution transformed traditional agriculture with: (1) the application of farm machinery for land preparation, reaping, hauling, irrigating, land clearing, fertilizer, manure and pesticide application; (2) the development and application of fertilizers, insecticides and weedicides; (3) application of sophisticated irrigation systems; (4) the application of principles of genetics to plant and animal breeding and (5) the practice of monoculture. These technologies have led to staggering increases in crop and animal production and productivity, larger farms and fewer farms and farmers [1-2, 6] and increased negative impact on the sustenance base [1-2, 6-8]. Another phase of agricultural evolution involved the application of information technologies, biotechnologies and modern science-based business management practices to organize and operate food production systems, leading to further gains in efficiency and productivity. Striking features of this phase include the following: large corporate style farms, drastic decline in family farms and profound innovations in the application of biotechnologies to the improvement of plants and animals. The progressive evolution of man's food gathering and food production relationship with his sustenance base (the ecology or environment) is characterized by: (1) his increasing capacity to apply science in developing the technologies used to manipulate the sustenance base or the ecological capital to meet his needs for food and fiber; and (2) the progressive ecological impact of these technologies. Prior to the phase of intensive application of science to agriculture, food production could be described as nature-based with food production and population more or less in balance with nature.

2. The impact of agriculture on the environment[2]

Rachel Carson's seminal work "Silent Spring" documented the environmental impact of insecticide on the environment [9]. Other authors including [1-2, 6-8] have documented an increasing environmental impact of conventional industrial agricultural technologies. Among the major impacts are point and non-point pollution from fertilizers and pesticides use; deforestation; desertification; salinization; soil erosion and sediment deposition downstream; degradation of water aquifers, accumulation of toxic compounds, loss of biodiversity; and

---

2 This section of the chapter is drawn extensively on the work of [4-5].

habitat fragmentation. The net effect of these impacts over time will be to reduce the capacity of the sustenance base to support increases in food production to meet the needs of future generations and the needs of those who currently suffer from hunger and malnutrition.

These concerns regarding health, as well as the environmental impacts and sustainability of conventional industrial agriculture have led to efforts directed at developing more sustainable alternatives as described by [10-13]. Alternatives, variously described as organic food production systems, community supported agriculture (CSA), community-based agriculture, and civic agriculture have begun to resonate and garner significant public support. These alternative approaches to food production are community-based food production systems. Community-based agriculture initiatives are nature-based and produce food in an environmentally sustainable manner [14-15]. Sustainable agricultural production systems practice crop rotation, no-till farming, diverse cropping patterns, use of organic matter or organically derived fertilizers, integrated pest management, biological control, cover cropping, timing of planting, leaving land in fallow, a variety of water conservation techniques and make optimum use of the natural biological cycles. The objective of a sustainable agricultural system is to forge a symbiotic relationship with the ecological capital and in the process learn to use the resources it provides without affecting the capacity of the ecological capital to support food production. This approach is tantamount to using a portion of the interest from an investment portfolio and ploughing back some earnings to ensure the continued productive capacity of the base investment capital. In contrast, conventional industrial agriculture views the ecology as primary capital input or raw material that is to be manipulated or consumed in the production process. The focus of sustainability in food production is to develop a food production system that mirrors or integrates with the natural ecology in which it exists. It is believed that such a system would achieve the highest degree of sustainability--the capacity to persist through time as a system of food production.

3. Sustainable agriculture the undergirding principle of organic agriculture[3]

What exactly is sustainable agriculture? Scholars and technocrats alike don't agree on a single definition, primarily because: (1) there is no way a single definition of the concept could be applied to cover the diversity of ecologies, cultural and economic conditions under which agriculture is practiced, and (2) there are several stakeholders, with a vested interest in the concept, who cannot agree on a single definition [16]. Essentially then, the practice of sustainable agriculture will be defined by local ecological and social conditions. Even though there is lack of agreement on a single definition of sustainable agriculture, there is general agreement that conventional agriculture or industrial agriculture is not sustainable for reasons mentioned above. For example, conventional agriculture depends increasingly on energy supplies from nonrenewable sources, depends on a narrow genetic base and intensive use of chemical

---

3 This section of the chapter is drawn extensively on the work of [4-5].

fertilizers and pesticides. In addition, it relies on subsidies and price support, has an increasing negative impact on the environment as evidenced by the loss of species, habitat destruction, soil depletion, consumption of fossil fuels and water-use at unsustainable rates, and contributes to air and water pollution and risks to human health [17].

Notwithstanding the difficulties involved in defining sustainable agriculture, given the threat posed by conventional agriculture, scholars still continue to work to define and clarify the concept. For example, Ikerd [18] proposed the following definition: "...capable of maintaining its productivity and usefulness to society over the long run...it must be environmentally-sound, resource conserving, economically viable and socially supportive, and commercially competitive" (p.30). In a later work Ikerd argued that sustainability should be thought of as a goal to be achieved rather than a static concept with a fixed definition. Even though Ikerd's view has considerable intuitive appeal, we believe that having a working definition clarifies what a concept represents and provides the information needed for identifying its constituent elements and distinguishing it from other concepts. Description of an object or thing provides insight into the nature of what that thing is and what it can do. Since what a thing can do depends on what it is, insights into its nature enables us to hypothesize about potential courses of action regarding that thing. Or, put another way, insights developed from clarifying the definition of a sustainable agricultural production system enables us to design courses of action to attain a sustainable food production system.

In this chapter, we draw on Ikerd's definition and the definition of sustainable development proposed by [19]. We define a sustainable agricultural production system as the practice of agriculture to produce food and fiber that meets the needs of the current population without compromising the capacity of the ecological capital, on which it depends, to support the needs of future populations. This means the nutritional, recreational and fiber needs of current populations must be met within the ecological limits of our natural resource base (ecological capital). The primary elements making up our definition are: (1) need, (2) time, (3) ecological capital, (4) equity, (5) population and (6) practice. From our perspective, the first element, "need" entails consuming resources to satisfy a physiological or physical requirement over time. Technically, a need is a necessity that is not satisfied in a single instance; it is a continuing requirement. In this sense, a sustainable agricultural system is one that is capable of persisting through time to meet current and future needs. The second element, "time" is a key concept, because sustaining anything means making sure that the particular thing persists through time. In the case of a sustainable agricultural system, it means managing our relationship with the ecological capital in such a manner that it will continue to meet our needs and the needs of future generations. The third element in our definition, "ecological capital," represents the resource base or the stock of natural assets that support life and food and fiber production. Our definition of ecological capital varies slightly from that offered by [1]. In our definition, we emphasize the biological base (the ecosystem) from which all natural services and goods are derived. Wright [1], on the other hand, defines it as the sum of goods and services provided by natural and managed ecosystems (agriculture) that are essential to human life and well-being. We chose to use the ecosystem or biological base because if the ecosystem is degraded or depreciated, its productive capacity and ability to support food production through a managed ecosystem (agriculture) will be much reduced.

The fourth element, equity, refers to the necessity to manage the endowment of ecological capital to meet the needs of the current generation without damaging its capacity to provide for future generations. In the context of our definition, the principle of equity also implies observing rules of fairness in the production, distribution and marketing of food and in exploiting other goods and services provided by our endowment of ecological capital. Population, the fifth element, refers to the current generation who consumes the goods and services produced from ecological capital, as well as future generations who will be consuming future products and services from the ecological capital. The attainment of a sustainable agricultural production system depends on the size of the population whose needs are to be met, the consumption level of the population, and the type of technology used in the production process. The final element, practice, deals with not only the technology employed in the production process but also the political, economic and social factors that impinge on and shape the sustainable agricultural production system. Given our definition, the question becomes: what insights for action can we draw? From our perspective, there are four primary insights (our illustrations below draw on the work of [1]): First, the population or people whose needs are to be met by a sustainable agricultural production system may be viewed from a dual perspective. People are the beneficiaries of a sustainable agricultural production system. Second, people are agents who must be proactive in defining what a sustainable food production system should be.

If a sustainable food production system is to be more than a theoretical abstraction, agents-the beneficiaries-must be able to operationalize the system to produce sustainable benefits. In operationalizing the concept of a sustainable agricultural production system, both values and knowledge play a central role in this process. Knowledge tells us about the ecosystem and how it supports agricultural production and what sort of sustainable development is possible, while our system of values guides us in making a choice once our options have been made clear. In this sense, moving from abstraction to implementation will be guided by the process illustrated in Figure 1 below. As illustrated in Figure 1, a sustainable food production system must be economically feasible "meaning such a system must be affordable and economically efficient. The sustainable food production system must also be socially desirable "indicating that it must be in sync with the cultural disposition and values of the agents or people it will serve. Consistent with this view, [17], reject approaches to sustainability that focus on the description and development of sustainable farming practices regardless of the socio-productive characteristics of the farming systems in which they are applied. Finally, a sustainable food production system must be in harmony with the ecology which supports it. If the food production system is discordant with, or in any way detrimental to the ecology that supports it, such a food system will not be sustainable.

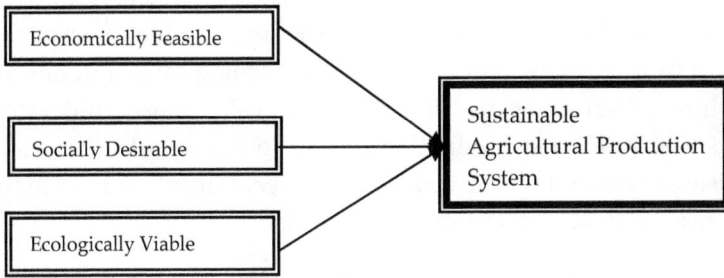

Figure 1. Sustainable food production system (adopted from [1])

4. Community and sustainable systems

Third, to make a food production system sustainable following the precepts depicted in Figure 1, the agents of such a system must act according to the framework illustrated in Figure 2. This is the point where community plays a vital role in crafting and managing a food production system to achieve sustainable objectives.

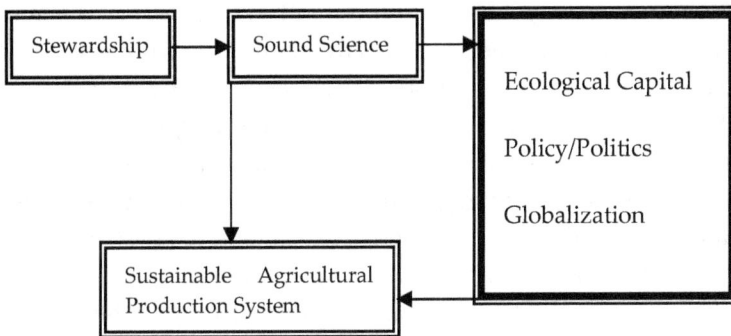

Figure 2. Framework for achieving a sustainable food production system (adopted from [1])

In Figure 2, stewardship entails employing ethical principles and values in choosing how sustainability is achieved. For example, sound-science provides knowledge about the ecosystem and the possibilities for supporting agricultural pursuits in a sustainable manner. It also informs us about how to make good decisions through policies and the political process. Science generates knowledge about specific sustainable practices and their efficacy. It tells us about the impact of globalization on the distribution of food, trade, and the spread of pollutants and diseases. In sum, science tells us what is and what is not possible. Good stewards must apply ethical standards and values to choose from among the possibilities that science generates in designing and implementing a sustainable agricultural production system, and in evaluating and adjusting the system to meet sustainable objectives. So then, the pivotal question becomes: Who gets to choose from among the possibilities that science generates?

Since food production in a sustainable system is inextricably linked to the local environment and the community's social and political infrastructure in which it exists, it follows that sustainable agricultural practices are defined by local ecological conditions and by the local social infrastructure which gives rise to the ethical values that guide stewardship. The connection of a sustainable food production system to ecological and social environments means that decisions concerning the design and development of sustainable agricultural production systems will have implications for everyone.

As a result, there will be several stakeholders with a vested interest in shaping the practice of sustainable agriculture. The reality is that citizens living in the same information rich environment as their leaders realize that the institutionalized bulwarks of authority are not omnipotent and that leaders are more or less ordinary people. Consequently, they assign less significance to the guidance of their leaders and institutions and have opted to become more reflective, proactive and self-regulating [20]. Implementing a sustainable agricultural production system in this context calls for collective action, because reflective and proactive citizens will insist on participating in the decision-making process. The support of diverse, reflective and proactive stakeholders is critical for ensuring that the values of stakeholders are reflected in defining and supporting the practice of sustainable agriculture.

Fourth, given that food systems depend on a healthy base of ecological capital regardless of their production technique, the sustainability of food systems can be conceptualized as existing on a continuum based on the level of integration with the natural ecosystem and the social environment in which it exists. At the high end of the continuum would be a production system that achieves the highest level of integration with the ecology and the social system in which it exists. And at the low end would be conventional/industrial agriculture. As indicated earlier, a sustainable system makes judicious use of available ecological capital by making optimal use of: biological cycles, the practice crop rotation, no-till farming, diverse cropping patterns, the use of organic matter or organically derived fertilizers, integrated pest management, biological control, cover cropping, timing of planting, leaving land in fallow and a variety of water and soil conservation techniques. To be sustainable, the food production system, as discussed earlier, must meet social and economic objectives within the limits of the ecology in which it exists. Sustainable food production must involve the community as consumers and stewards of the food production system. The system must also nurture and expand understanding of the interdependence of food production and the ecology which supports it. Considering that people are the agents and beneficiaries of a sustainable food system, communities must understand and accept that natural resources are finite, recognize the limits on economic growth, and encourage equity in resource allocation [17]. In other words, the drive for economic efficiency must be tempered by the need to preserve ecological capital and ensure social and economic equity. The trend toward large-scale Industrial profit driven farming has implications for the economic health of rural communities. For example, studies have demonstrated that independent hog farmers generate more jobs, more local retail spending, and more local per capita income than do larger corporate operations. Comparisons between conventional industrial agriculture and sustainable systems indicate that organic agriculture and sustainable systems are productive and economically competitive [17].

Given the concept of sustainable food system describe herein, we suggest that sustainable food systems exist on a continuum. The top end of the continuum would define a food production system that is nature-based and which achieves the highest level of integration with the ecology and social system in which it exits. We would label this highly ecologically and socially integrated food production system organic agriculture. Our conception of organic agriculture presented here is consistent with the definition proposed by Codex Alimentarius Commission which states that:

"Organic agriculture is a holistic production management system which promotes and enhances agro-ecosystems health including biodiversity, biological cycles, and soil biological activity. It emphasizes the use of management practices in preference to the use of off-farm inputs, taking into account that regional conditions require locally adapted systems. This is accomplished by using, where possible, cultural, biological, and mechanical methods, as opposed to using synthetic materials, to fulfill any specific function within the system." (Quoted in [21] pp.6)

At the low end of the continuum, displaying the lowest level of integration would be conventional industrial agriculture. Between these two extremes would be food production systems that manifest varying degrees of ecological and social integration or levels of sustainability. So then, organic agriculture is the ideal that we should work toward achieving as we strive to achieve a sustainable food system.

In today's market place there is a growing demand for organic products. And consumers seem willing to pay a premium price for products carrying organic quality labels. Questions that arise are how reliable are these quality labels and what level of confidence should consumers put in such labels? Usually the control process is carried out by independent certifiers who are guided by criteria promulgated by rule-making agencies. Certifiers must be vigilant and succeed in revealing departures from standards and opportunistic behavior in order for quality assurance labels to build up the reputation necessary to serve as a reliable quality signal. However, in the case of Potemkin attributes (where the desirable attribute is based on a process such as in organic production) there is the potential for quality statements to be made with little risk of disclosure of departures from standards, because consumer agencies, NGOs, and public authorities are usually not able to verify marketing claims or discover opportunistic behavior. What is needed to deter opportunistic behavior and identify departures from accepted standards is a quality monitoring protocol that covers the whole supply chain and ensures on-site inspections throughout the production process [22]. Another approach is to ensure stricter audit standards and rigorous training of certifiers, but these approaches are likely to increase the cost of certification and the resultant cost of organic products, which will drive down demand for products that are already offered to consumers at premium prices. In our view, a less expensive, organically-based and a more resilient approach would entail shortening the supply chain and fostering closer connection between producers and consumers. We envisage that the community and farmers would fulfill the role of active co-stewards (the community of consumers and producers) of the organic food production system. As co-stewards of an organic food production system, farmers and consumers would be organized in networks that exchange ideas, share experiences and information and work together to solve problems. In this situation, an effective self-monitoring protocol that is grounded in a culture of trust and commitment to standards could emerge. The opportunity for farmers and consumers to interact as co-stewards would create an appreciation for the attributes that consumers' value, the relationship between these valued attributes, the production process and the price farmers are able to fetch for their product. On the other hand, consumers would get an appreciation for the process that produces the valued attribute. Over time, the "deep trust" that would develop between producers and consumers as a result of the co-creation of understanding of the role of consumer and farmer in meeting each other's need would lead to an effective monitoring system. This level of understanding could potentially lead to the identification of points of weakness in the process; whereupon, co-stewards would take action to modify existing protocols that would reduce the likelihood of opportunistic behavior.

The idea of entrepreneurial social capital espoused by [23] provides a conceptual basis for our proposed co-creation of an effective and inexpensive monitoring system. In the instance outlined above, co-stewards (the community of consumers and producers) have the potential to serve as a catalyst for mobilizing entrepreneurial social infrastructure (ESI). [23] Defines ESI as having three elements: symbolic diversity, resource mobilization and quality of networks. Symbolic diversity enables co-stewards to encourage participation, dissent, accept challenges to the status quo and embrace constructive controversy and critiques; it encourages people to focus on the process and the arguments instead of the personalities involved. It also encourages resource mobilization, which involves promoting local investment by residents in the community, equity in resource and risk distribution and collective investment in the community. Quality networks are encouraged by establishing horizontal and vertical linkages. Horizontal networking links co-stewards in similar circumstances and promotes learning by sharing experiences and information from different perspectives. Vertical linkages draw on resources of others operating in dissimilar circumstances, or in different systems. *It enables co-stewards to attract resources from private and public sources outside the community, for example, from entities with different levels of expertise and capacity relevant to the problem at hand* (our emphasis).

5. Assessing consumers preferences toward production system and consumers preferences for the attributes of fresh fruits and vegetables[4]

This next section will examine the attitude of consumers toward organic, sustainable and conventional production system and consumers preferences for the attributes of fresh fruits and vegetables. As discussed earlier, sustainable production lies between organic and conventional production system on our continuum described above. Thus, a sustainable agricultural production system is operationalized as employing good agricultural practices (judicious use of synthetic fertilizers and pesticides), integrated pest management and emphasizes the use of natural cultural practices and fertilizers and insecticides from natural sources as much as possible.

---

4 This section of the chapter is drawn extensively on the work of [24-25].

6. Measuring preferences for food production systems

The advent of specialized stores offering organic produce and products and the allocation of supermarket self-space to organic produce and products attest to the increasing demand for food and food products produced under alternative production systems. The emergence of alternative food production systems and the discussion in the public domain concerning the health, environmental and social benefits they offer vis a vis conventional production systems may have, at the very least, sensitized consumers about the opportunities that exist for making food purchasing decisions based on the type of production system and its perceived benefits. Additionally, the promotion of healthy eating habits and the need for increased consumption of fruits and vegetables [26-28], plus the well-publicized need for environmental conservation [19] amplify the salience and relevance of differences between the food production systems in terms of their health, environmental and socio-economic impact. Consequently, our objective here is to assess consumer attitudes toward food produced under the following food production systems – conventional agriculture, sustainable alternatives and organic along five criteria – contribution to environmental conservation, food safety, food quality, contribution to wellness and contribution to community economic development by using Analytic Hierarchy Process (AHP).

6.1. Data and methodology

The sample was designed following the protocol described by [29]. It was drawn proportionate to population size by county in Georgia, North Carolina and South Carolina. After specifying the sampling frame parameter, the required sample was purchased from Survey Sampling Inc. Data were collected from a random sample of 252 respondents, which represents a cooperation rate of 30 percent. Researchers designed and formatted an analytic hierarchy questionnaire to collect data via a telephone survey. Enumerators asked consumers to compare three food production systems: conventional, sustainable and organic in terms of which consumers would prefer farmers to use in producing the fresh fruits and vegetables that they purchase or consume; taking into consideration environmental, food safety, food quality, wellness, and community development issues.

This study employed Analytic Hierarchy Process (AHP) to derive a measure of an individual consumer's preference for production systems in terms of the selected criteria which is consistent with previous research conducted in the U.S. [30]. The AHP, which was developed by [31], is one of the most commonly applied multi-criteria decision-making techniques. AHP is a subjective tool for analyzing qualitative criteria to generate priorities and preferences among decision alternatives (For more detailed information about AHP, see [32-34]. The AHP model, illustrated in Figure 3, was used to assess consumers' preferences for production systems in terms of environment, food safety, food quality, wellness and community development.

Cluster analysis was used to separate consumers into groups by: age, education and employment status. The aim of cluster analysis is to classify observations into relatively homogeneous groups called clusters, such that each cluster is as homogeneous as possible with respect to the clustering variables [35-36]. Researchers would then be able to determine if consumers' preferences for production systems varied by age, education or employment status. The Kolmogorov-Smirnov test was used to check whether the clustering variables were normally distributed and the Kruskal Wallis test was used to compare clusters.

Further analysis employing multidimensional scaling (MDS) was used to obtain "perceptual mapping of consumers' preferences for production systems. By transforming consumer judgments of overall preferences into distance represented in multidimensional space, MDS plots the three production systems and five criteria on a map such that those systems and criteria that are perceived to be very similar to each other are placed near each other on the map, and those systems and criteria that are perceived to be very different from each other are placed far away from each other on the map. In this way MDS provides a visual representation of the pattern of proximities (i.e., similarities) among the set of production system and the set of criteria employed in their assessment [36].

6.2. Results and discussion

Consumers were grouped into three clusters. The mean of the variables used in the analysis is presented by the clusters in Table 1. There were statistically significant differences among clusters on the variables age, education and employment. The mean age (40.85) is the lowest in Cluster 1 and the highest (80.35) in Cluster 3. Education level is the highest (4.94) in Cluster 1 and lowest in cluster 3 (2.41). Employment status changes from employed in Cluster 1 (2.13) to unemployed in Cluster 3 (2.97). Cluster 1 is labeled "Young professional", while the cluster 2 and cluster 3 are labeled "Older-technician" and "Oldest-unemployed" respectively.

Variables	Clusters			Kruskal Wallis Test	
	1	2	3	Chi-Square	Asymp. Sig
Age	40.85	63.48	80.35	191.962	0.000
Education*	4.94	4.32	2.41	29.596	0.000
Employment**	2.13	2.52	2.97	69.077	0.000

* 1: Less than high school, 11: Professional/doctorate degree; ** 1: Part time, 2: Full time, 3: Unemployed

Table 1. Cluster analysis by age, education and employment

Table 2 displays the number of consumers by the clusters. The data show that 52.1 percent of consumers are "young professional", 33.5 percent are "older-technician", while the "oldest-unemployed" accounts for 14.4 percent.

In the AHP Model, consumers were asked to assess conventional, sustainable and organic production systems, taking into account the ability of each to generate benefits related to environmental conservation, food safety, food quality, wellness and community economic development. The AHP model for assessing preferences for production systems in terms of

Clusters	Frequency	Percent	Cumulative Percent
Young professionals	123	52.1	52.1
Older technician	79	33.5	85.6
Oldest-unemployed	34	14.4	100.0
Total	236	100.0	

Table 2. Consumer distribution by clusters

these criteria is defined in Figure 3. The goal is to determine consumers' preferences for food produced under three production systems using the following criteria: environmental conservation, food safety, food quality, wellness and community economic development. These criteria are the perceived benefits generated by each system. In the AHP model illustrated below, consumers are being asked to choose their preferred food production system from among the alternatives: conventional, sustainable and organic production systems based on environmental conservation, food safety, food quality, wellness and community economic development criteria.

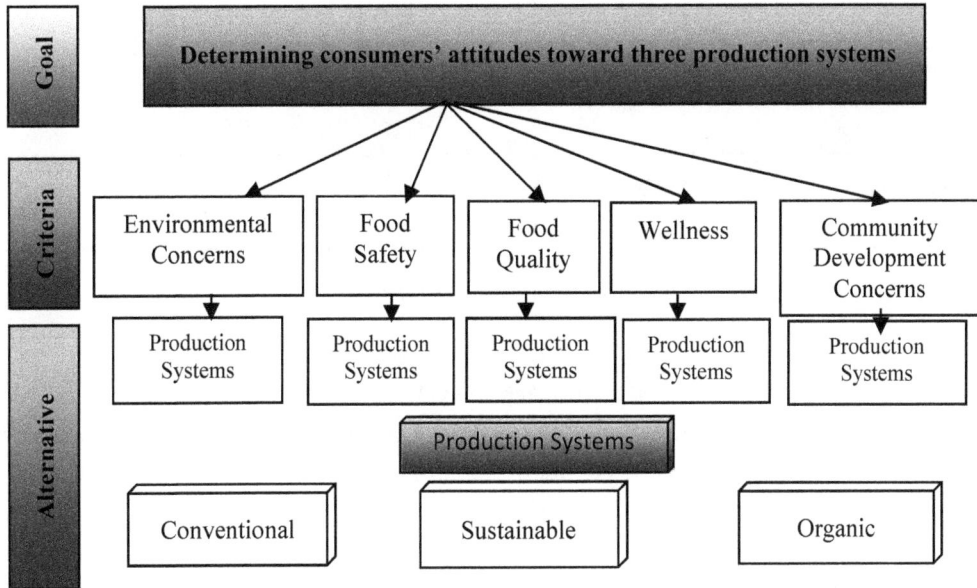

Figure 3. AHP model for consumer attitudes toward food production systems

Table 3 shows the results obtained by applying the AHP model. The last column in Table 3 indicates consumers' average priority ratings for each criterion. The results indicate that consumers accorded priority in the following order to food safety (0.281) followed by wellness

(0.275), food quality 0.209), environmental concerns (0.144) and community development concerns (0.091). Consumers considered food safety and wellness to be more important attributes or features of a food production system than other attributes such as food quality and the capacity of the food system to contribute to community development or environmental quality. In each row of Table 3, the preference scores for each type of production systems are presented. The third column of Table 3 shows that organic agriculture is preferred, when considered alone, based on its perceived capacity to generate benefits associated with wellness (0.575), food quality (0.533), safety (0.530), environmental concerns (0.515) and community development (0.514). The average preference rating of 0.544 shown in the last row of Table 3 indicates that consumers prefer the organic production system over the sustainable alternative and conventional agriculture, which were assigned preference ratings of 0.274 and 0.182 respectively.

Criteria	Conventional	Sustainable	Organic	Preference
Environmental Concerns	0.203	0.282	0.515	0.144
Food Safety	0.186	0.284	0.530	0.281
Food Quality	0.195	0.272	0.533	0.209
Wellness	0.162	0.262	0.575	0.275
Community Development Concerns	0.209	0.278	0.514	0.091
Final Decision	0.182	0.274	0.544	

[1] Consumer preference scores are ranged between 0 and 1.The sum of each row, excluding the preference in the last column, is equal to 1.00.

Table 3. Consumers' attitudes toward food production systems by the criteria

Since consumers' preferences for the production systems of food may be influenced by their demographic traits and behaviors [37], demographic traits may be used, where heterogeneity in consumers preferences exists, to segment consumers into groups based on their demographic characteristics. Cluster analysis was employed using the variables: age, education and employment status to identify discrete groups of consumers based on their preferences. The results indicate that there are three distinct groups of consumers: young professionals, older-technician and oldest-unemployed. The preference ratings each segment assigns to the three production systems are shown in Table 4. These results show that there were no statistically significant differences among the consumer segments in their preferences for the food production systems. Table 5 indicates priorities each segment assigned to criteria used to assess the food production systems; young professionals accorded a higher priority to community development concerns than the other two groups.

Production Systems	Clusters			Kruskal Wallis Test	
	Young professionals	Older technician	Oldest-unemployed	Chi-Square	Asymp. Sig
Conventional	0.170	0.188	0.211	1.287	0.526
Sustainable	0.274	0.284	0.253	2.264	0.322
Organic	0.556	0.528	0.536	1.242	0.537

Table 4. Consumer attitudes toward food production systems for each segment

Criteria	Clusters			Kruskal Wallis Test	
	Young professionals	Older technician	Oldest-unemployed	Chi-Square	Asymp. Sig
Environmental Concerns	0.142	0.158	0.132	1.783	0.410
Food Safety	0.270	0.281	0.309	4.569	0.102
Food Quality	0.199	0.222	0.208	2.493	0.288
Wellness	0.292	0.261	0.277	1.135	0.567
Community Development Concerns	0.097	0.078	0.073	5.273	0.072

Table 5. Consumer attitudes toward the criteria generated by production systems for each segment

Figure 4 shows the consumers' perceptual map derived from multidimensional scaling. The map illustrates the pattern of proximities for food production systems and the criteria consumers used to assign preference ratings. Kuskal's stress value was used to measure goodness-of-fit. The stress value is a number on a scale from 0 (perfect fit) to 100 (the map captures nothing about the data). In general, researchers are looking for a stress value less than 20 [38]. In the MDS results, Kruskal's stress value is 10 for this two dimensional model and $R^2=0.97$. Similar to factor analysis, there is a measure of difficulty in interpreting the conceptual mapping of consumers' perception. To overcome this difficulty, researchers rely on their knowledge of the subject, existing theory and plausible rationale along with the weights associated with the stimulus coordinates to make good sense of the derived stimulus configuration [39]. The results indicate that consumers view organic production systems as quite dissimilar to the other production systems. Additionally, organic production is perceived as being associated with food safety and wellness, but not with environmental and community development benefits. On the other hand, consumers perceive a sustainable system of production to be associated with environmental and community development and food quality. Consumers see conventional as being dissimilar to organic and sustainable production systems and not associated with environment, community development, food quality, food safety and wellness. Consequently, the y axis is labeled as environmental /community development and the x axis as conventional production system. This means that moving from left to right along the x axis the production system becomes more conventional, and moving along the y axis from top to bottom environmental sensitivity of the production system decreases.

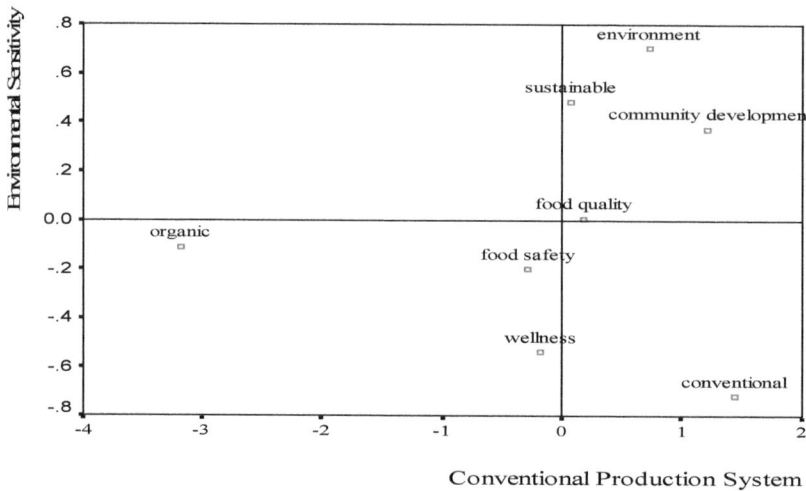

Figure 4. Perceptual mapping of consumers attitudes toward food production systems

6.3. Conclusion

Consumers accord the highest preference score to organic production, followed by sustainable and conventional production systems respectively. Moreover, in according higher priority to food safety and wellness, consumers appear to be more concerned with criteria that are more tangible in terms of the consequence for consumers' personal and immediate well-being. Since our findings indicate that consumers don't associate organic food production with benefits for environmental and community development, there is a need to design education programs that will convince consumers that there are socioeconomic and environmental benefits to be derived from organic production. However, education programs without community institutional support are not likely to succeed. Community members must be engaged as co-creators of initiatives that are intended to change attitudes and create awareness. We recall that proactive and reflective community members live in the same information rich environment as their leaders and those of us considered to be experts. Proactive and reflective citizens tend to assign less significance to leaders and experts, they insist on participating in the decision making process, they want to co-create programs that have implications for their livelihood. As a result, a truly sustainable food system (organic) must become embodied, and an intimate part of the lived experience of people and communities. After all, it is action that creates destiny. So if a sustainable food system is to become a part of our future, it has to become a way of life, and a pattern of living that is acted out as part of the everyday life story of communities [4].

7. Measuring preferences for food attributes

7.1. Data and methodology

The sample was designed following the protocol described by [29]. The sample was drawn proportionate to population size by county in Georgia, North Carolina and South Carolina. After specifying the sampling frame parameters, the required sample was purchased from Survey Sampling Inc. Researchers designed and formatted a Fuzzy Pair-wise Comparison (FPC) questionnaire to be compatible with the data collection protocol of Survey Monkey, and trained enumerators to use the questionnaire to collect the data. Enumerators asked consumers to make pair-wise comparisons of five food attributes: nutritional value, hygiene, taste, affordable price and freshness, in order to determine their preference for one attribute over the other. The selected attributes are consistent with the studies which have been done in the U.S. [30]. Data were collected from a random sample of 412 respondents.

In this study, FPC was used to derive a measure of an individual consumer's preferences for fresh fruit and vegetable attributes. The main reasons for using FPC are: 1) The FPC is similar to traditional pair-wise comparisons. Consumers are asked to compare the attributes one pair at a time. However, unlike the traditional pair-wise method, consumers are not forced to make a binary choice between two attributes. Consumers are permitted to indicate the degree of preference for one attribute over another, and response indicating indifference between attributes is permitted. 2) Unlike the other methods, the scale values are based on the respondent's entire set of paired comparisons. 3) FPC more accurately represents the natural range of response patterns that are possible. The consumer's fuzzy preference matrix R with elements can be constructed as follows [40]:

$$R_{ij} = \begin{cases} 0 & \text{if } i = j \, \forall i, j = 1,...,n \\ r_{ij} & \text{if } i \neq j \, \forall i, j = 1,...,n \end{cases} \tag{1}$$

In the FPC method, a measure of preference, μ can be calculated for each attribute by using the consumer's preference matrix R. The intensity of each preference is measured separately using the following equation:

$$\mu_j = 1 - \left(\sum_{i=1}^{n} R_{ij}^2 / (n-1) \right)^{1/2} \tag{2}$$

where μj has a range in the closed interval [0,1]. A larger value for μj indicates greater intensity of preference for attribute j. Consequently, fresh fruit and vegetable attributes are ranked from most to least preferable by evaluating the μ values. Then, Friedman and Kendall's W tests were used to evaluate the relative importance of attributes and the extent of agreement among consumers with respect to two or more rankings. In identifying consumer preferences, researchers ranked the importance of the attributes following [37].

Cluster analysis was used to separate consumers into groups using the variables: age, education and employment status. Cluster analysis is a technique used for combining observations or objects (answer, person, opinion, etc.) into groups or clusters. The aim of cluster analysis is to classify observations into relatively homogeneous groups called clusters such that each cluster is as homogeneous as possible with respect to the clustering variables [35-36]. The Kolmogorov-Smirnov normality test was used to check whether the clustering variables showed normal distribution, and then the Kruskal Wallis test was used to compare different groups of clusters.

Multidimensional Scaling (MDS) was used to obtain a perceptual mapping of consumers' preferences for fresh fruit and vegetable attributes. Given a matrix of perceived similarities between attributes of fresh fruit and vegetables, MDS plots the attributes on a map such that those attributes that are perceived to be very similar to each other are placed near each other on the map, and those attributes that are perceived to be very different from each other are placed far away from each other on the map.

7.2. Results and discussion

In this study, consumers were grouped into three clusters. The mean of the variables used in the analysis is presented by clusters in Table 6. There were statistically significant differences among clusters on the variables; age, education and employment of consumers in the sample. The mean age (37.19) is the lowest in Cluster 1 and the highest (77.54) in Cluster 3. Education level is the highest (5.63) in Cluster 1, whereas Cluster 3 has the lowest level (3.87). Employment status changes from employed in Cluster 1 (2.06) to unemployed in Cluster 3 (2.89). Therefore, cluster 1 is labeled "Young professional", while the cluster 2 and cluster 3 are labeled "older-employed" and "oldest-unemployed", respectively.

Variables	Clusters			Kruskal Wallis Test	
	1	2	3	Chi-Square	Asymp. Sig
Age	37.19	58.15	77.54	339.960	0.000
Education[+]	5.63	4.57	3.87	24.101	0.000
Employment[++]	2.06	2.33	2.89	92.656	0.000

[+]1: Less than high school, 11: Professional/doctorate degree;+[+]1: Part time, 2: Full time, 3: Unemployed

Table 6. Cluster analysis by age, education and employment

Table 7 indicates the number of consumers by clusters. The data show that 47.7 percent of consumers are "older-employed worker", whereas 34.8 percent are "young professional", while 17.5 percent represent "oldest-unemployed".

Clusters	Frequency	Percent	Cumulative Percent
Young professional	141	34.8	34.8
Older-employed worker	193	47.7	82.5
Oldest-unemployed	71	17.5	100.0
Total	405	100.0	

Table 7. Consumer distribution by clusters

Descriptive statistics for consumers' pair-wise comparisons of the attributes of fresh fruit and vegetables obtained from the FPC model are presented in Table 8. The fresh fruit and vegetable attributes are ranked from most to least preferable using the reported degree of the consumers' preferences. The results show that the fresh fruit and vegetable attribute most preferred by consumers is freshness with a preference rating of 0.579. Gao, et al. [37] reported a similar pattern of preference in their study on consumer preferences for fresh citrus. Consumers prefer the other food attributes in the following order: taste (0.452), hygiene (0.449), nutritional value (0.428) and affordable price (0.411). In this sample, consumers seem to value freshness, taste and hygiene over price and nutritional value. The Friedman test was used to see if there was a difference in the rankings of the fresh fruit and vegetable attributes.

Attributes	Mean	Standard deviation	Minimum	Maximum
Nutrition Value	0.428	0.122	0.024	0.929
Hygiene	0.449	0.142	0.049	1.000
Taste	0.452	0.128	0.049	0.868
Affordable Price	0.411	0.154	0.000	0.735
Freshness	0.579	0.159	0.150	1.000

Significant by Friedman test for $p < 0.01$; Kendall's W=0.11

Table 8. Descriptive statistics of consumer preferences towards fresh fruits and vegetable attributes

The Friedman test, which is significant ($\chi^2 = 177.71$; $p < 0.01$), confirms that some attributes are preferred over the others. Kendall's W test was used to measure the degree of agreement among consumers. The value of Kendall's W is 0.11, which indicates that the level of agreement among consumers in ranking the attributes is very low. A low level of agreement among consumers is an indication of the heterogeneity of consumers' preferences for the attributes of fresh fruits and vegetables.

Since consumers' preferences for the attributes of fruits and vegetables may be influenced by their demographic traits and behaviors [37], demographic traits may be used, where hetero-geneity in consumers preferences exists, to segment consumers into groups based on their demographic characteristics. The present study employed cluster analysis using the variables: age, education and employment status to identify discrete groups of consumers based on their preferences. The results indicate that there are three distinct groups of consumers: young professionals, older-employed worker and oldest-unemployed. The results also showed that there was a statistically significant difference among the groups in their preferences for the freshness attribute of fruits and vegetables. Young professionals accorded a higher priority to freshness than the other two groups (Table 9).

Variables	Clusters			Kruskal Wallis Test	
	Young professional	Older worker	Oldest-unemployed	Chi-Square	Asymp. Sig
Nutrition Value	0.414	0.434	0.440	2.980	0.225
Hygiene	0.449	0.449	0.452	0.104	0.949
Taste	0.440	0.456	0.473	1.860	0.395
Affordable Price	0.395	0.413	0.436	1.909	0.385
Freshness	0.598	0.579	0.547	6.027	0.049

Table 9. Consumer preferences for fresh fruits and vegetable attributes by clusters

Figure 5 shows consumers' perceptual map with attribute positioning derived from multidi-mensional scaling (MDS) analysis of consumers' preferences for the attributes of fresh fruits and vegetables. In the MDS results, Kruskal's STRESS measure is 0.03863. A satisfactory measure should be less than 0.05 for a two dimensional model [39]. $R^2=0.99404$ shows that the model's goodness-of-fit is perfect. The analysis indicates that consumers perceive freshness as a distinct food attribute, which is quite separate from taste, hygiene, nutritional value and affordable price. On the other hand, consumers do not seem to perceive hygiene and nutritional value as distinct attributes, that is, consumers tend to accord the same level of priority to hygiene and nutritional value. Similarly, consumers tend to accord the same level of priority to taste and price.

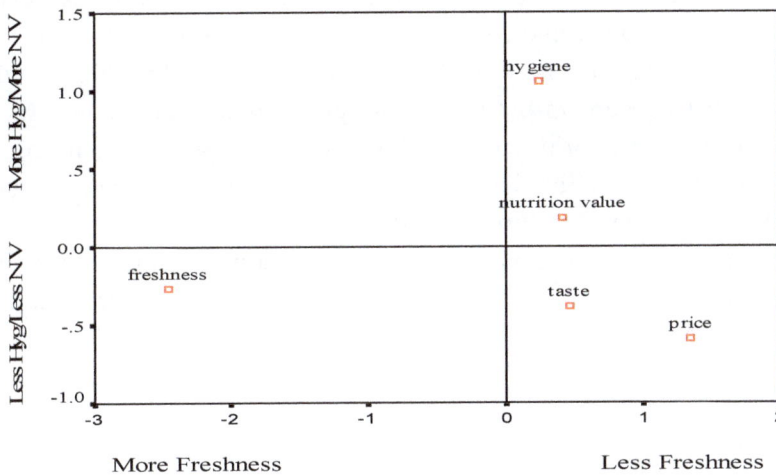

Figure 5. Perceptual mapping of consumers preferences for fresh fruit and vegetable attributes.

7.3. Conclusions

Consumers in making purchasing decisions pay more attention to freshness, taste and hygiene attributes of fresh fruits and vegetables than they do price and nutritional value, when these attributes are considered individually. However, multidimensional scaling shows that consumers tend to associate taste and price when making purchasing decisions, which may explain consumers' love for inexpensive tasty fast food, especially in the case of low income consumers. These results indicate that consumers may not be using all the information available in selecting which food to purchase based on the preference ratings. Therefore, the need exists to educate consumers on the connection among the food attributes and their relevance to healthy eating habits and a healthier lifestyle, particularly the nutritional value attribute. Knowledge about the subgroups of consumers – young professionals, older-employed and oldest-unemployed –provides a basis for farmers, especially farmers supplying urban and suburban farmers' markets, to tailor their products based on the needs of these groups of consumers, a strategy known as market segmentation. For example, the results indicate that the priority or preference of young professionals is for freshness. Extension should use this information to assist farmers to select and display their produce to promote freshness in order to sell more to the higher income young professionals. In summary, these results present extension with an opportunity to (1) assist farmers in marketing their produce in order to meet the needs of specific groups of consumers and (2) in developing a holistic education program, that teaches consumers to use information available on all the attributes: price, taste, hygiene and nutritional value in making purchasing decisions.

In sum, studies have shown that organic farming delivers more environmental benefits, in particular, it delivers more ecosystem services than conventional agriculture [41]. Additionally, contingent on the crop, soil and weather conditions, yield from organic agriculture is equal to that from conventional systems [42]. In the context of a sustainable food production system, organic agriculture goes further in meeting the condition of ecological feasibility (Fig

1), and the evidence seems to indicate that, with further advances in the development of organic technologies, it will become economically feasible. In terms of the third condition to be met-being socially acceptable-in striving for overall sustainability (Fig 1), evidence from our work shows that consumers prefer organic production systems over the alternative systems. Thus far, the future of organic production systems seems promising, but further research is needed to advance the development of organic technologies, disseminate these technologies, increase supply to reduce cost and make organic products affordable to a wider range of consumers, formulate supporting policies, and educate consumers on the value of organic food production systems in contributing to a sustainable food production system.

Author details

Terrence Thomas[1*] and Cihat Gunden[2]

*Address all correspondence to: twthomas@ncat.edu

1 North Carolina Agricultural and Technical State University, Department of Agribusiness, Applied Economics and Agriscience Education, USA

2 Ege University, Faculty of Agriculture, Department of Agricultural Economics, Izmir, Turkey

References

[1] Wright R.T. Environment Science: Toward a Sustainable Future, 9th Edition. Upper River, NJ: Pearson Prentice Hall; 2005.

[2] Wright R.T. Environment Science: Toward a Sustainable Future, 10th Edition. Upper River, NJ: Pearson Prentice Hall; 2008.

[3] Kaufmann R. Cleveland C. Environmental Science, 1st Edition. Boston: McGraw Hill; 2008. (Book)

[4] Thomas T., Yeboah O., Bukenya J., Gray B., Ofori-Boadu V. Accounting for Socio-Cultural Factors in Designing Sustainable Agricultural Production Systems. Journal of Environmental Monitoring and Restoration 2007; 3(1) 127-139.

[5] Thomas T., Yeboah O., Ofori-Boadu V., Fosu E. Assessing Local Community Support for Sustainable Agricultural Production Systems. Journal of Environmental Monitoring and Restoration 2008; 5 191-203.

[6] Raven P., Berg L. Environment 5th Edition. NJ: John Wiley & Sons; 2006.

[7] Botkin D., Keller E. Environmental Science: Earth as a Living Planet, 5th Edition. NJ: John Wiley & Sons; 2006.

[8] Ikerd J. Small Farms: The Foundation for Long-Run Security. Paper Presented at "A Time to ACT: Providing Educators with Resources to Address Small Farm Issues" sponsored by University of Illinois, Agroecology/Sustainable Agriculture Program, Effingham and Peoria, IL; 2002. http://www.ssu.missouri.edu/faculty/jikerd/papers/I11Small.html

[9] Carson R. Silent Spring. NY: Fawcett World Library; 1962.

[10] Delind L.B. Place, Work, and Civic Agriculture: Fields for Cultivation. Agriculture and Human Values 2002; 19 217-224.

[11] Lapping M.B. Big Places, Big Plans. In Furuseth, O. (ed). Perspectives on Rural Policy and Planning. Hampshire: Ashgate Publishing Limited; 2004.

[12] Lyson T. A., Guptill A. Commodity Agriculture, Civic Agriculture and the Future of U.S. Farming. Rural Sociology 2004; 69 370-385.

[13] Flora C.B. Sustainability of Agriculture and Rural Communities. In Francis, C. A., Flora C.B., King, L.D. (eds). Sustainable Agriculture in Temperate Zones. NY: John Wiley & Sons; 1990.

[14] Cone C.A., Myhre A. Community Supported Agriculture: a Sustainable Alternative to Industrial Agriculture? Human Organization 2000; 59(2) 187-196.

[15] Lamb G. Community Supported Agriculture: Can it Become the Basis of a New Associative Economy? The Threefold Review 1994; 11 39-44.

[16] Rigby D., Caceres D. Organic Farming and the Sustainability of Agricultural Systems. Agricultural Systems 2001; 68 21-40.

[17] Horrigan L., Lawrence R.S., Walker P. How Sustainable Agriculture can Address the Environmental and Human Health Harms of Industrial Agriculture. Environmental Health Prospective 2002; 110(5) 445-456.

[18] Ikerd J. The Need for a Systems Approach to Sustainable Agriculture. Agriculture, Ecosystems & Environment 1993; 46 147-160.

[19] World Commission on Environment and Development. From One Earth to One World: An Overview. Oxford: Oxford University Press; 1987.

[20] Snyder, David Pearce. 2007. Five Meta-Trends Changing the World. In Jackson, Robert M. (ed). Annual Editions: Global Issues 06/07. Dubuque, IA: McGraw-Hill; 2007.

[21] Borron, S. Building Resilience for an Unpredictable Future: How Organic Agriculture can Help Farmers Adapt to Climate Change. Rome: Food and Agriculture Organization of the United Nations; 2006.

[22] Jahn G., Schramm M., Spiller A. The Reliability of Certification: Quality Labels as a Consumer Policy Tool. Journal of Consumer Policy 2005; 28(1) 53–73.

[23] Flora C.B., Flora J.L. Entrepreneurial Social Infrastructure: a Necessary Ingredient. The Annuals of the American Academy of Political and Social Science 1993; 529 48-58.

[24] Gunden C., Thomas T. Assessing Consumer Attitudes towards Fresh Fruit and Vegetable Attributes. Journal of Food Agriculture & Environment 2012; 10(2) 132-135.

[25] Thomas T., Gunden C. Investigating Consumer Attitudes toward Food Produced via Three Production Systems: Conventional, Sustainable and Organic. Journal of Food Agriculture & Environment 2012; 10(2) 132-135.

[26] Stewart H., Harris J.M. Obstacles to Overcome in Promoting Dietary Variety: The Case of Vegetables. Review of Agricultural Economics 2004; 27 21-36.

[27] U.S. Department of Agriculture. Increasing Fruit and Vegetables Consumption through the USDA Nutrition Assistance Programs. Food and Nutrition Service Progress Report, VA; 2008.

[28] Food and Agricultural Organization (FAO) Increasing Fruit and Vegetable Consumption Becomes a Global Priority: 2003. http://www.fao.org/english/newsroom/focus/2003/fruitveg1.htm.

[29] Dillman D., Smyth J., Christian L. Internet, Mail, and Mixed Mode Surveys: The Total Design Method, 3rd Edition. NJ: John Wiley & Sons; 2009.

[30] Moser R., Raffaelli R., Thilmany-McFadden D. Consumer Preferences for Fruit and Vegetables with Credence-Based Attributes: A Review. International Food and Agribusiness Management Review 2011; 14 121-142.

[31] Saaty T.L. The Analytic Hierarchy Process: Planning, Priority Setting, Resources Allocation. NY: McGraw-Hill; 1980.

[32] Saaty T.L. Fundamentals of Decision Making and Priority Theory. Pittsburgh: RWS Publications; 2006.

[33] Saaty, T.L. 2008. Decision Making for Leaders. RWS Publications, Pittsburgh.

[34] Saaty T.L., Kirti P. Group Decision Making: Drawing Out and Reconciling Differences. Pittsburgh: RWS Publications; 2008.

[35] Tabachnick B.G., Fidell L.S. Using Multivariate Statistics, 5th Edition. NY: Pearson; 2007.

[36] Hair J.F., Black W.C., Babin B.J., Anderson R.E. Multivariate Data Analysis, 7th Edition. NJ: Prentice Hall; 2010.

[37] Goa Z., House L.O., Gmitter F.G., Valim M.F., Plotto A., Baldwin E.A. Consumer Preferences for Fresh Citrus: Impacts of Demographic and Behavioral Characteristics. International Food and Agribusiness Management Review 2011; 14 23-40.

[38] Johnson K. Qualitative Methods in Linguistics. Oxford: Blackwell Publishing; 2008.

[39] Mazzocchi M. Statistics for Marketing and Consumer Research. London: SAGE Publications; 2008.

[40] Van Kooten G.C., Schoney R.A., Hayward K.A. An Alternative Approach to the Evaluation of Goal Hierarchies among Farmers. Western Journal of Agricultural Economics 1986; 11 40-49.

[41] Sandhu, H. S., Wratten, S. D., & Cullen R. Organic Agriculture and Ecosystem Services. Environmental Science & Policy 2010; 13 1-7.

[42] Pimentel, D., Hepperly, P., Hanson J., Douds, D., & Seidel, R. Environmental, Energetic, and Economic Comparisons of Organic and Conventional Farming Systems. BioScience 2005; 55 (7) 573-582.

Lithuanian Organic Agriculture in the Context of European Union

Vytautas Pilipavičius and Alvydas Grigaliūnas

1. Introduction

Organic agriculture is a production system that sustains the health of soils, ecosystems and people. It relies on ecological processes, biodiversity and cycles adapted to local conditions, rather than the use of inputs with adverse effects. Organic agriculture combines tradition, innovation and science to benefit the shared environment and promote fair relationships and a good quality of life for all involved (As approved by the IFOAM General Assembly in Vignola, Italy in June 2008) [1]. The growing criticism of intensive agricultural practices that leads to a deterioration of natural resources and a decrease of biodiversity has progressively led to more environmental constraints being put on agricultural activities through an "ecologization" of agricultural policies [2]. Organic agriculture nowadays is well accepted in governmental and scientific institutions, and organic products are highly appreciated by consumers [3]. Organic agriculture can contribute to solving the food crisis and mitigating global climate change as long as it is based on the principles of agroecology. However, organic agriculture must also be integrated with certain conventional agricultural practices in order to maintain rational production and satisfy the food requirements of the population [4].

Organic agriculture is the form of farming that sees nature as a living organism. Determination of further health and productivity of this organism differs from that in conventional agriculture. It is a versatile system that supplements and conditions the environment friendly measures enabling regulation of ecological system. At the same time application of chemical synthetically measures can be abandoned. Ecological system must be maintained, preserved and rebuilt as completely as possible as we are absolutely dependent on its functioning. This trend of farming requires general systematic way of thinking – the course of the entire process should be considered when applying each measure [5]. The goal of organic farming is to give priority to long-term ecological health, such as biodiversity and soil quality, rather than short-term productivity gains [6, 7]. In low potential agricultural areas characterized by soil degradation and erosion, organic agriculture can provide a means to break the downward spiral of resource degradation and poverty [8]. Organic farming represents an innovation in agriculture that is both lauded and deplored. Agricultural innovations are accepted on four broad levels: research, extension, farmer and community (not necessarily in that order) [7]. The implementation of European legislation as well as various national pesticide action plans and public policies pertaining to organic agriculture, could bring about major changes to agricultural practices within the coming years [2].

Farming is only considered to be organic at the EU level if it complies with Council Regulation (EC) No 834/2007 [9] amended in (EU) No 1030/2013 [10], which has set up a comprehensive framework for the organic production of crops and livestock and for the labelling, processing and marketing of organic products, while also governing imports of organic products into the EU. The detailed rules for the implementation of this Regulation are laid down in Commission Regulation (EC) No 889/2008 [11] amended in commission implementing regulation (EU) No 392/2013 [12].

Many of the environmental problems of great concern today are either directly or indirectly related to past and present agricultural practices. The only way to preserve the nature and especially agrocenosis for future generations in the XXI century is organic sustainable development.

2. Material and methods

The area of study is Lithuanian organic agriculture in the context of European Union that Lithuania joined in 2004.

Lithuanian territory situated between 56°27′N and 53°53′N latitude, 20°56′E and 26°50′E longitude occupies intermediate geographical position between west Europe oceanic climate and Eurasian continental climate. Climate of the Lithuanian territory forms in different radiation and circulation conditions. Differences in these conditions hardly cross the boundaries of microclimatic differences; therefore, Lithuania belongs to western region of the Atlantic Ocean continental climatic area [13, 14] with average annual precipitation of 675 mm (572-907 mm) and temperature of 6-7 °C.

A review of the scientific literature on organic agriculture and the research on the development and perspectives of Lithuanian organic agriculture are evaluated. Statistical information is an important tool for understanding and quantifying the impact of political decisions in a specific territory or region [15].

In the manuscript statistical data of organic agriculture of Lithuania and European Union are analysed scientifically and analytically with interpretation of praxis.

3. Development of organic agriculture

To make spread and development of alternative agriculture in the world easier the International Federation of Organic Agriculture Movement IFOAM was established in 1972. This federation unites majority of alternative agriculture movements in different states. At present IFOAM includes over 700 organizations of alternative-ecological agriculture from more than 100 countries. Several directions of IFOAM activity cover consulting, data exchange and standardization [5].

The area of certified organic agricultural land in the world continuously is tending to increase (Figure 1). During ten years period from 1999 to 2009 organic agricultural area increased by 3.4 times and reached 37.2 million hectares.

Sharp development of organic agriculture lightly moved to more balanced development while increase in organic area from 2008 to 2009 already consisted 5.7%. Such dynamics is in conformity with the organic area development in Europe (Figure 2) and separate countries (Figure 6), as well as in Lithuania (Figure 7).

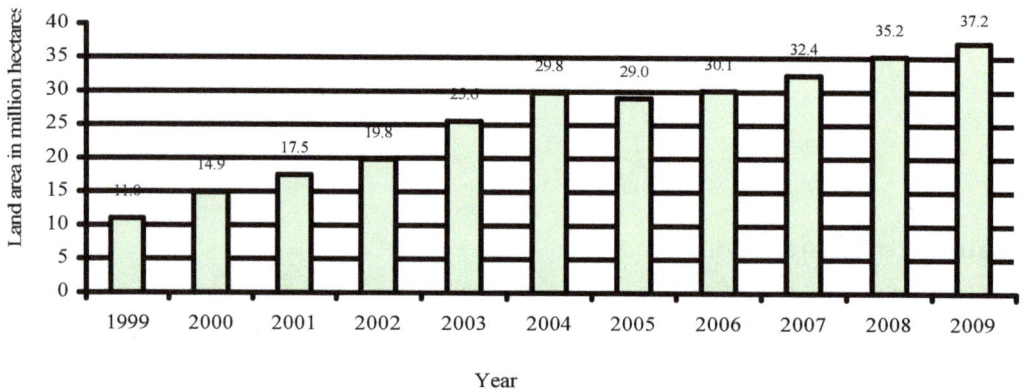

Figure 1. Development of organic agricultural land 1999-2009 in the world [16]; Source: FiBL, IFOAM and SOEL 2000-2011; www.fibl.org

Area of certified organic agricultural land in Europe increased by million hectares or 12 percent from 2008 to 2009 (Figure 2) while in early years (2005, 2006, 2007, 2008) increase consisted of 6.4-7.3%. The organic area 2009 in Europe covered 9.3 million hectares while European Union countries covered 8.4 million hectares from them. The geographical distribution of area and share of organic agriculture in Europe by country in 2009 is presented in figure 3. The biggest organic areas were established in Spain, Italy and Germany.

Geographical distribution of fully converted organic crop area in 2012 in EU is presented in figure 4 and share of organic agriculture in percent (Figure 5). It showed that the highest certified organic agricultural areas are in Italy, Spain, Poland, France and the United Kingdom. However, share of total utilized agricultural area occupied by organic farming in Italy, Spain, Poland, France and the United Kingdom were 8.9%, 7.5%, 4.6%, 3.6% and

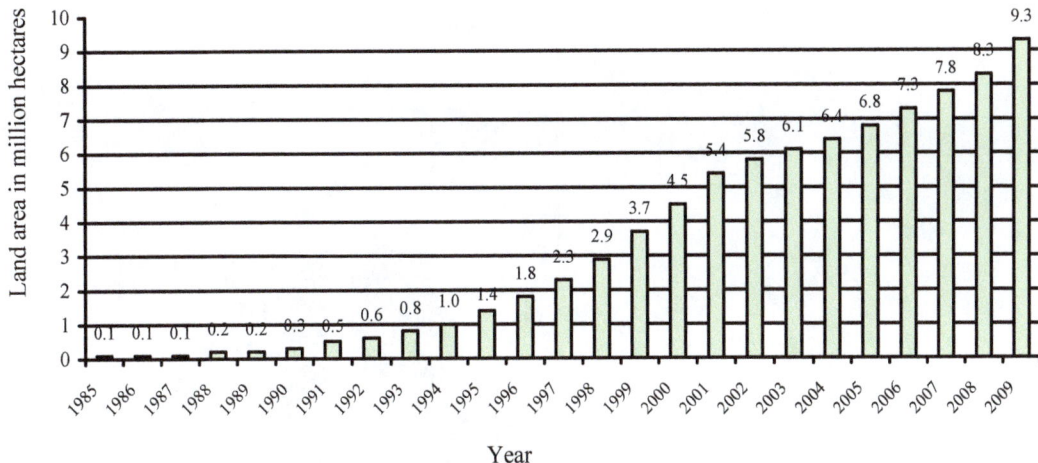

Figure 2. Development of the organic agricultural land in Europe 1985-2009 [16]; Source: FiBL, Aberystwyth University, AMI/ZMP

3.4% accordingly (Figure 6). The highest share from total agricultural land area in 2012 were in Austria 18.6%, Sweden 15.8%, Estonia 14.9% and Czech Republic 13.1%. The least managed organic areas were fixed in Malta and Bulgaria (till 1%). EU-28 average covered 5.7% in 2012 (Figure 6).

Lithuanian organic agriculture area development was extremely rapid from the start of organic agriculture as a farming form in 1990 [17]. From 2004 till 2012 area of organic agriculture in Lithuania increased 3.8 times till 162655 hectares in 2012 (Figure 7). Contrary to continuously increase of organic area the organic farm number has tendency to decrease from reached maximum 2855 farms in 2007 while 2511 farms left in 2012. Therefore, the average size of certified organic farm in Lithuania constantly increased and in 2012 already covered 64 ha (Figure 8). The distribution of certified organic farms in Lithuania according to the farm size in hectares is presented in figure 9. The most organic farms (47%) operated area is till 30 ha. Organic farms with 31-50 ha, 51-100 ha and 101-300 ha covered 16.5%, 17.0% and 16.5% respectively. The biggest organic farms with the area above 300 hectares make 2.5% of the general number of organic farms in Lithuania (Figure 9).

The average size of organic agricultural holdings in 2007 was 37 ha for the EU-27 as a whole, compared to 13 ha for all agricultural holdings [19]. The average size of each agricultural holding (farm) in the EU-28 was 14.2 hectares in 2010 [20]. In general, the average size of holdings in the organic sector was larger in most of the Member States and smaller only in Denmark, France and Luxembourg. The most noticeable differences were seen in the Czech Republic (223 ha compared to 89 ha) and Slovakia (421 ha compared to 28 ha), (Figure 10). One possible reason for these sometimes big differences is the use of a more extensive method of farming within the organic sector [19].

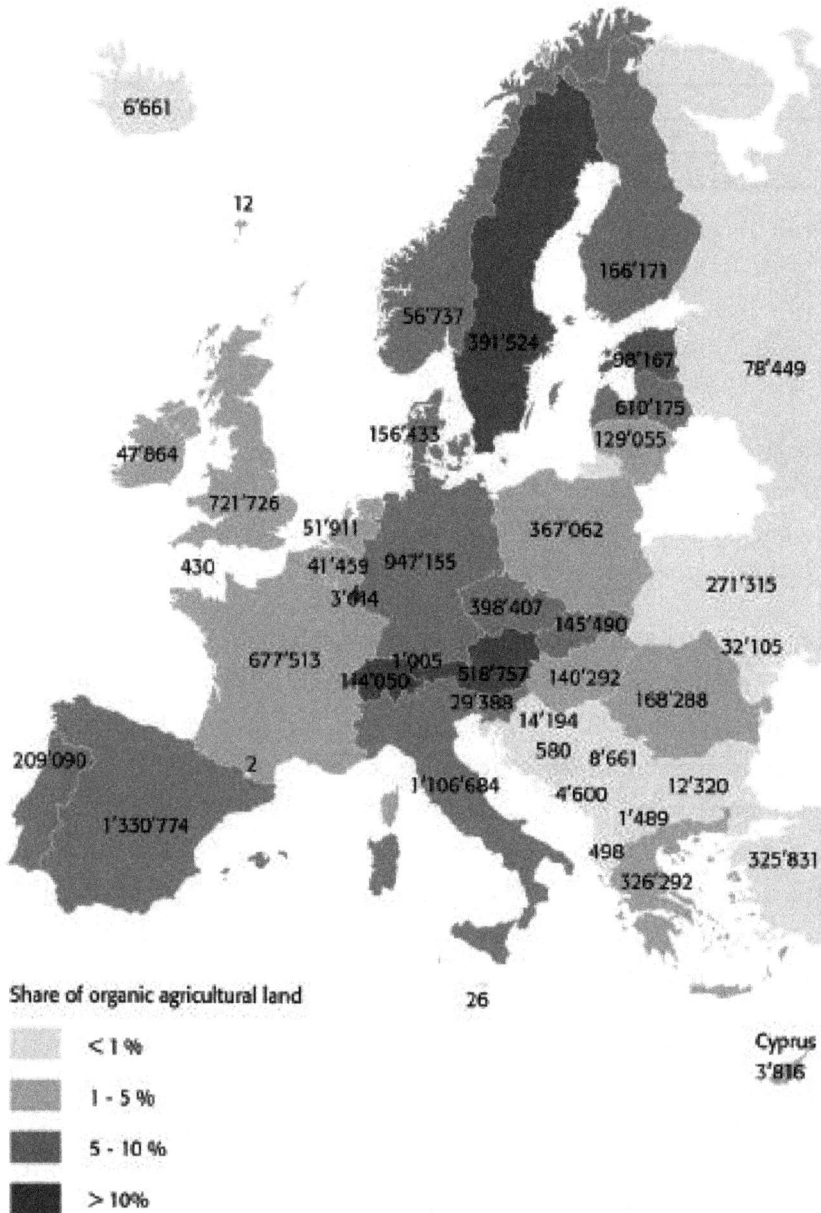

6'661

12

166'171

56'737

391'524

98'167   78'449

610'175

156'433   129'055

47'864

721'726

51'911   367'062

430   41'459 947'155   271'315

3'014

398'407

145'490   32'105

677'513   1'005

144'050 518'757 140'292

29'388   168'288

14'194

580   8'661

209'090   2   12'320

1'106'684   4'600   1'489

1'330'774   498   325'831

326'292

Share of organic agricultural land   26

< 1 %   Cyprus
                                          3'816
1 - 5 %

5 - 10 %

> 10%

Figure 3. Organic agricultural land by country in Europe 2009 [16]; Source: FiBL Survey 2011; www.fibl.org

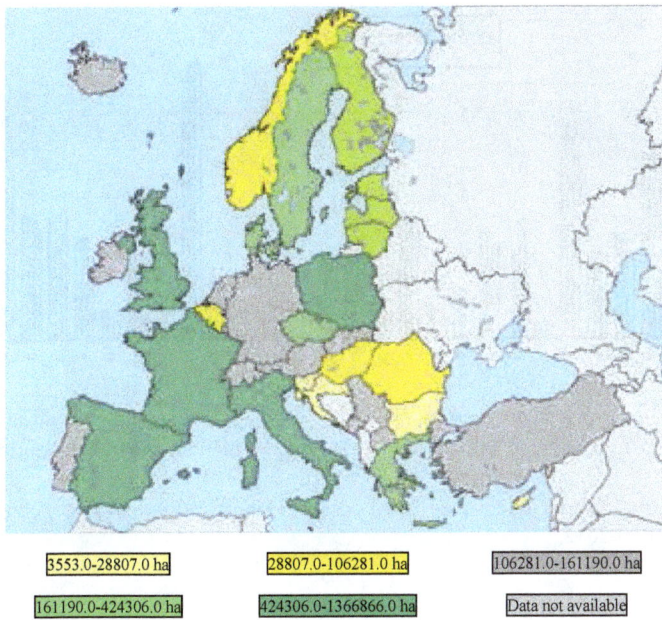

Legend:
- 3553.0-28807.0 ha
- 28807.0-106281.0 ha
- 106281.0-161190.0 ha
- 161190.0-424306.0 ha
- 424306.0-1366866.0 ha
- Data not available

Source: Eurostat (Available from: <http://epp.eurostat.ec.europa.eu/tgm/mapToolClosed.do?tab=map&init=1&plugin=0&language=en&pcode=tag00098&toolbox=types>)

Figure 4. Organic crop area (fully converted area) in hectares in European Union 2012

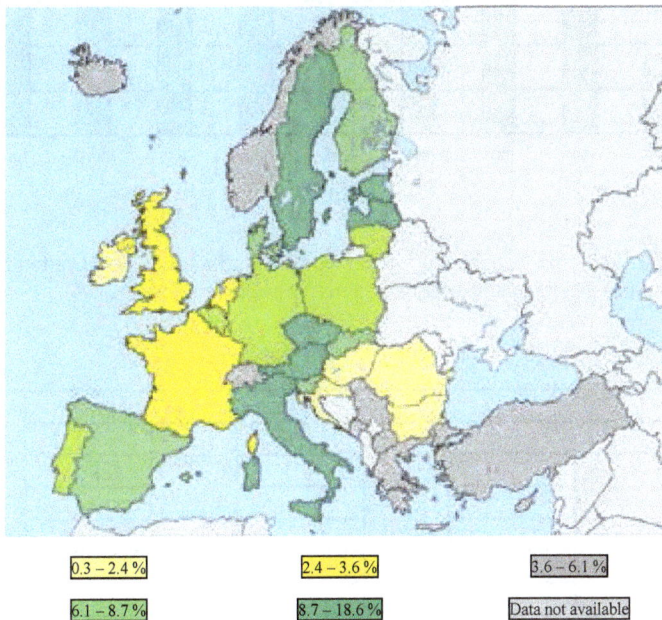

Legend:
- 0.3 – 2.4 %
- 2.4 – 3.6 %
- 3.6 – 6.1 %
- 6.1 – 8.7 %
- 8.7 – 18.6 %
- Data not available

Source: Eurostat (Available from: <http://epp.eurostat.ec.europa.eu/tgm/mapToolClosed.do?tab=map&init=1&plugin=0&language=en&pcode=tsdpc440&toolbox=legend>)

Figure 5. Share (%) of agricultural area under organic farming in European Union, 2012.

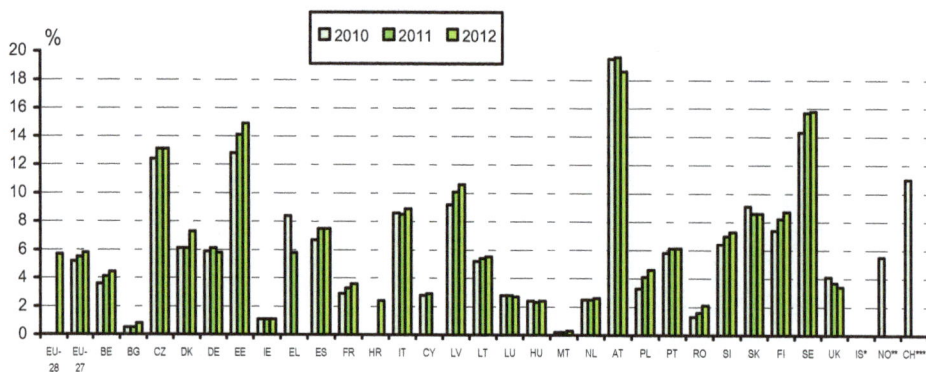

Note. *-data not available, **-2009, ***-2005; Source: Eurostat (Available from: <http://epp.eurostat.ec.europa.eu/tgm/table.do?tab=table&init=1&plugin=0&language=en&pcode=tsdpc440>)

Figure 6. Share of total utilized agricultural area (UAA) occupied by organic farming (fully converted and under conversion) in per cent (%) 2010-2012.

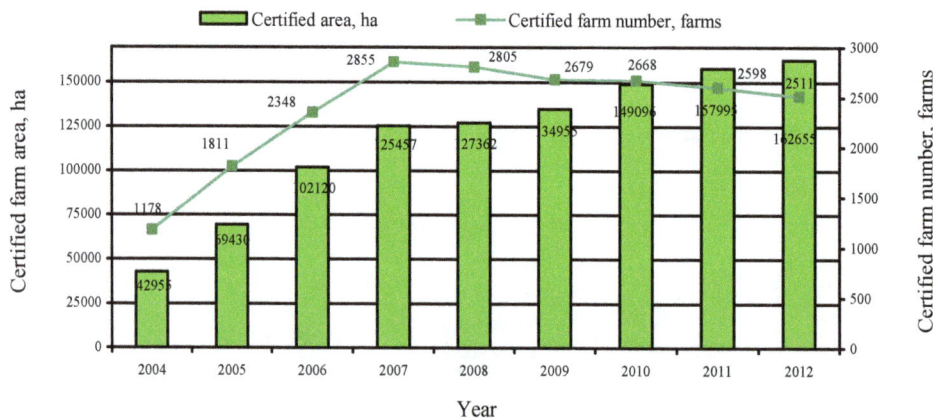

Figure 7. Dynamics of certified organic agriculture farms number and area (fully converted and under conversion) during 2004-2012 in Lithuania including area of fishery farms [18]; Source: Ekoagros.

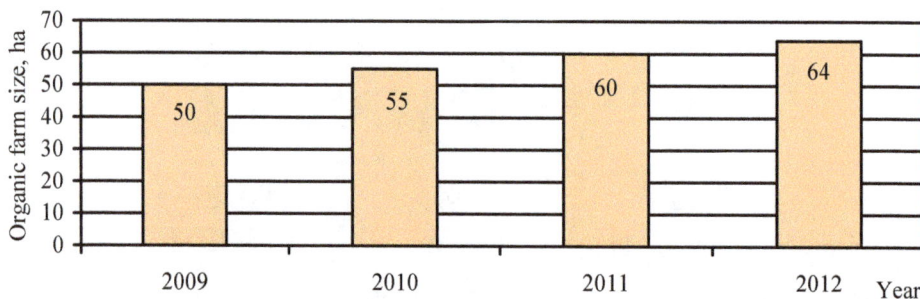

Figure 8. Average size of certified organic farms in Lithuania, including area of fishery farms [18]; Source: Ekoagros

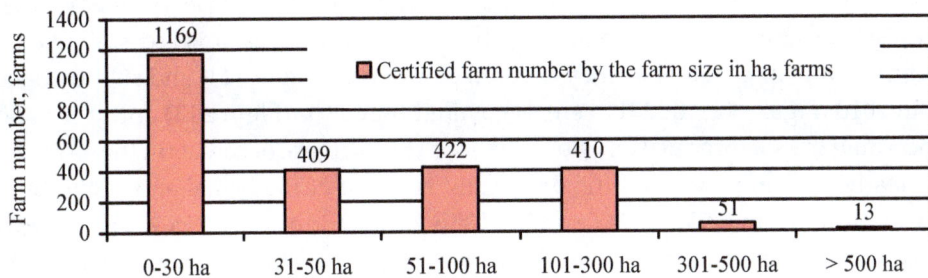

Figure 9. Differentiation of certified organic farms in Lithuania according to the farm size in hectares 2012 [18]; Source: Ekoagros.

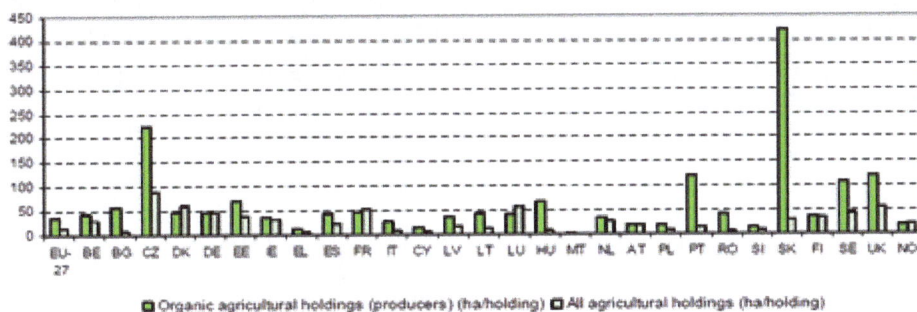

Note. Organic area: DK, MT data 2006; Organic holdings: MT data 2006; PL data 2008 [19];

Source: Eurostat (food_act2, food_in_porg1, ef_ov_kvaaesu)

Figure 10. Average size of agricultural holdings/farms in European Union, 2007 (ha/holding)

4. Organic crop production

4.1. Dynamics of organic agriculture crop structure and distribution

The structure of agriculture in the Member States of the European Union varies as a function of differences in geology, topography, climate and natural resources, as well as the diversity of regional activities, infrastructure and social customs. There were 12.2 million farms across the EU-28 in 2010, working 174.1 million hectares of land (the utilised agricultural area) or two fifths (40.0 %) of the total land area of the EU-28 [20]. Farming land covers nearly 54% of the total area of Lithuania, with arable land and grassland accounting for 70% and 27% respectively.

The organic crop area structure and its dynamics in Lithuania from 2008 to 2012 are presented in figures 11, 12, 13 and 14. The main crop by the occupied area covered soil productivity exhausting cereals (spica cereals). They covered 49% in 2008 (Figure 11) and slightly decreased till 46% in 2010 (Figure 12). In 2011 cereals covered only 36% (Figure 13) and it was positive turn concerning the soil productivity preservation. The decrease of cereals in the crop structure in 2011 mostly was influenced by the drastically increase of medicinal and potherbs area in the crop structure from 0.55% in 2008 to 5.01% in 2010 and to 15.94% in 2011. The reason of such rapid increase in organic medicinal herb area could be 3 times higher subsides than for organic cereals (see table 6). However, difficulties in growing and processing of medicinal and potherbs as well still limited local market for medicinal and potherbs because of much higher price turn its area back to 4.12% in 2012 (Figure 14). Parallel area of the cereals in organic crop structure increased back till 43%. Vegetables are still not common in Lithuanian organic agriculture with range of 0.38%-0.30% in crop structure; at the European level organically grown vegetables unfortunately take analogous position (Figure 15). The permanent crops in Lithuania have tendency to decrease from 4.3% (2008) to 3.26% (2012) as orchards (from 1.02% to 0.74%) and small-fruit plants (from 3.28% to 2.52%) while in Europe organic permanent crops in 2009 increased to 11% (Figure 15) compared with 2008 (Figure 16).

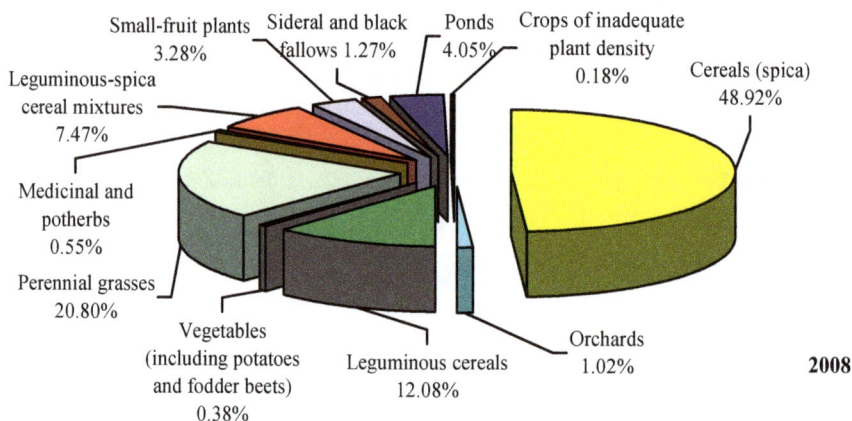

Figure 11. Organic crop area structure including fallows and ponds in Lithuania 2008. Source: Ekoagros; www.ekoagros.lt

The next very important component in the organic crop structure is permanent grassland. At the European level organic permanent grassland takes fine 46% (Figure 15) while in Lithuania 26% in 2012 (Figure 14) that show slow but continuously increase from 20.8% in 2008 (Figure 11 and 16). To control organic agriculture as producing system it was introduced requirements of minimal plant density in the crop which is much lower than optimal crop density. Therefore, crops of inadequate plant density cover very low area – only 0.18% – 0.48% (Figure 11-13) and come to praxis exceptionally because of very unfavourable meteorological conditions of the year or season. It helps to prevent organic agriculture from unfair farming as well subsidies for such areas are suspended.

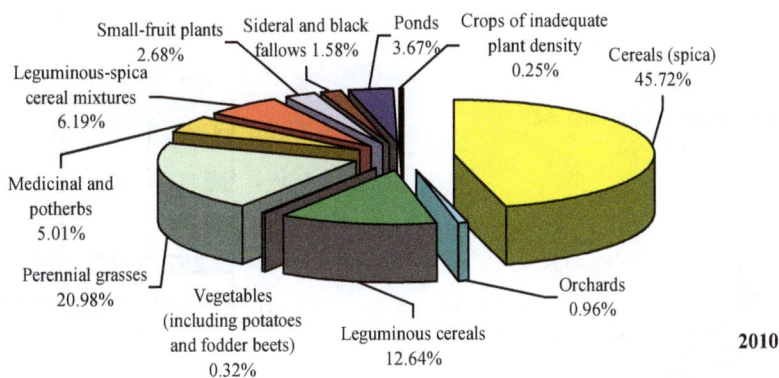

Figure 12. Organic crop area structure including fallows and ponds in Lithuania 2010. Source: Ekoagros; www.ekoagros.lt

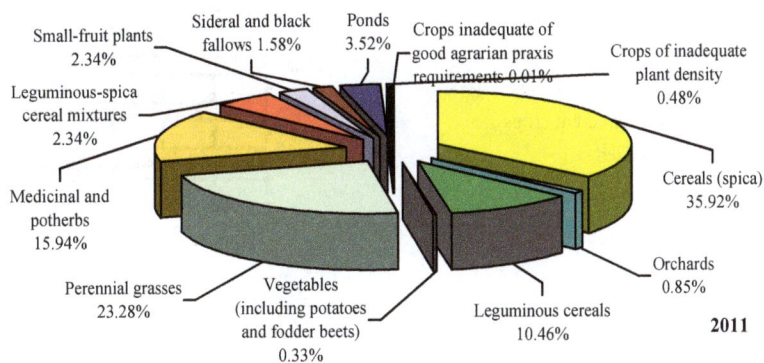

Figure 13. Organic crop area structure including fallows and ponds in Lithuania 2011. Source: Ekoagros; www.ekoagros.lt

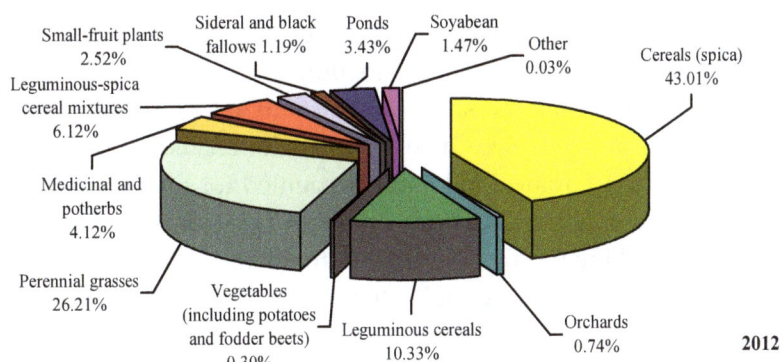

Figure 14. Organic crop area structure including fallows and ponds in Lithuania 2012 [18]. Source: Ekoagros

Distribution of main agricultural land use
types 2009

The main arable crops 2009

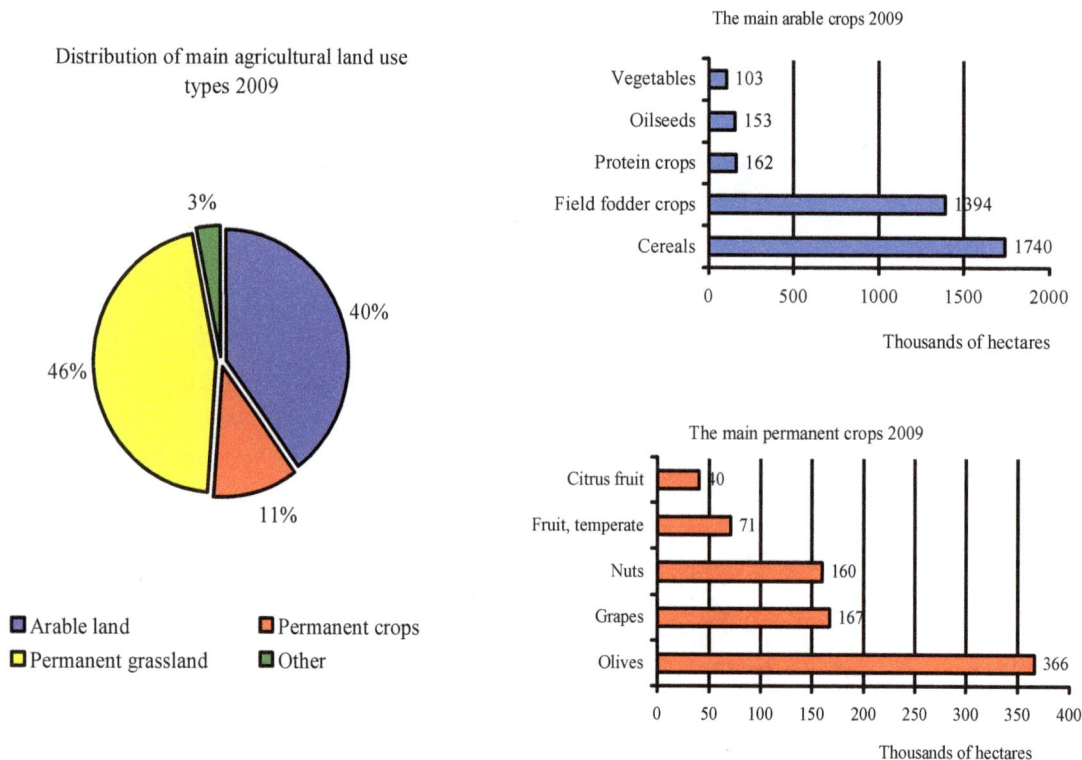

■ Arable land   ■ Permanent crops
□ Permanent grassland   ■ Other

The main permanent crops 2009

Figure 15. Use of organic agricultural land in Europe 2009 [16]; Source: FiBL Survey 2011; www.fibl.org)

The three main crop types grown organically are arable land crops (mainly cereals, fresh vegetables, green fodder and industrial crops), permanent crops (mainly fruit trees and berries, olive groves and vineyards) and pastures and meadow land [19]. In most of the Member States and Norway, permanent crops account for a relatively insignificant share of the fully converted area of these three main types (less than 5%). In 2008, permanent crops accounted for between 10% and 20% in Bulgaria, Denmark, Greece, Poland and Portugal, while in Spain and Italy the share was over 20%. Cyprus and Malta were in the lead with 41% and 80% respectively. Olive trees predominated in both countries. In 11 countries (including Norway) arable land crops accounted for the largest share of the land area (> 50%), while in 15 countries pastures and meadows predominated (>50%). Arable crops were significant in Finland and Norway with shares of 98% and 80% respectively (75% in Lithuania), while the Czech Republic (92%), Ireland (96%) and Slovenia (89%) were in the lead in terms of pastures and meadows (Figure 16) [19].

The most common crops in Lithuanian organic agriculture are cereals and in farms with animal husbandry-perennial grasses. In organic farms without animal husbandry perennial grasses as a matter of routine are absent as in this case the subsides for the perennial grasses are not

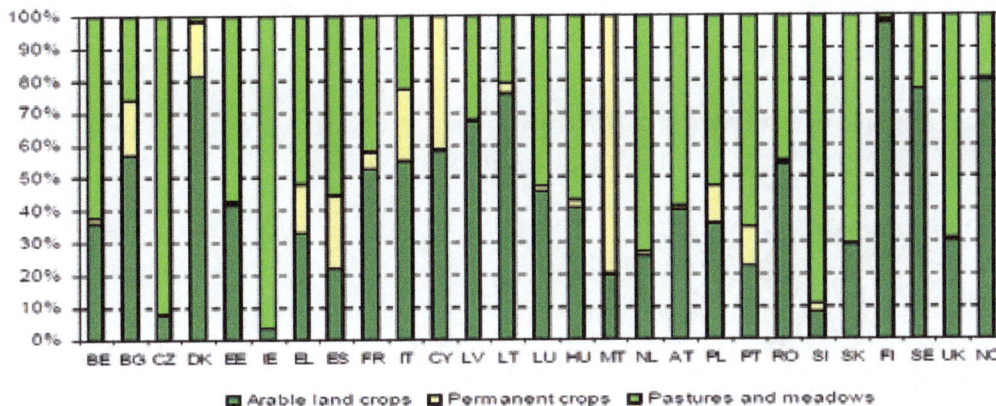

Note. MT data 2006, CY, PT data 2007; DE no data available; IE, ES share in total organic area, Source: Eurostat (food_in_porg1)

Figure 16. Share in percent (%) of arable land crops, permanent crops, pastures and meadows in fully converted area, 2008 [19].

paid. The distribution of wheat, barley and perennial grasses in Lithuania by the municipalities (2005) in comparison of organic to conventional agriculture are presented in figures 17, 18 and 19. To follow the relevance to grow wheat, barley and perennial grasses in separate territories are presented the land productivity points. Organic wheat covered 0.7%-3.1% and conventional 4.2%-8.7% in the crop structure in the most favourable land to wheat growing by the land productivity points (Figure 17). On the lands of 40-48 points of land productivity wheat covers 0.4%-12.9% and 0.3%-7.9% in organic and conventional agriculture accordingly. Higher share of wheat in conventional agriculture on the best lands could be explained by the higher share of cereals in the crop rotation of conventional agriculture and not seldom used sowing wheats after wheats. Though, sugar beets (that are not grown at all organically in Lithuania) take the first position on the best lands of conventional agriculture because of the highest profitability. On land of lower productivity, i.e. 32-40 points and less than 32 points organic wheat covers 0.1%-4.8% and 0%-2.7% while conventional wheat covers 0.3%-1.9% and 0%-0.5% accordingly (Figure 17).

Organic barley was the most common on 40-48 points land productivity reaching till 16.2% (2005) when conventional varied in the range of 1.5%-6.3% in the crop structure (Figure 18). On the best soils organic barley covered 0.5%-1.8% while conventional one took their highest share of 3.2%-6.8% in the crop structure.

Organic perennial grasses covered till 9.3% on land of 32-40 point productivity, till 5.7% on land of 40-48 point productivity and till 5.9% on land till 32 point productivity (Figure 19). On the best soils organic perennial grasses covered only from 0% till 0.9% area in the crop structure while conventional ones from 0.9% till 3%. On the land with decreasing land productivity from 40-48 to 32-40 and less than 32 points conventional perennial grasses covered till 3.6%, 6.3% and 8.2% accordingly (Figure 19).

Organic wheat area (ha and %) distribution in Lithuania by the
municipalities, 2005

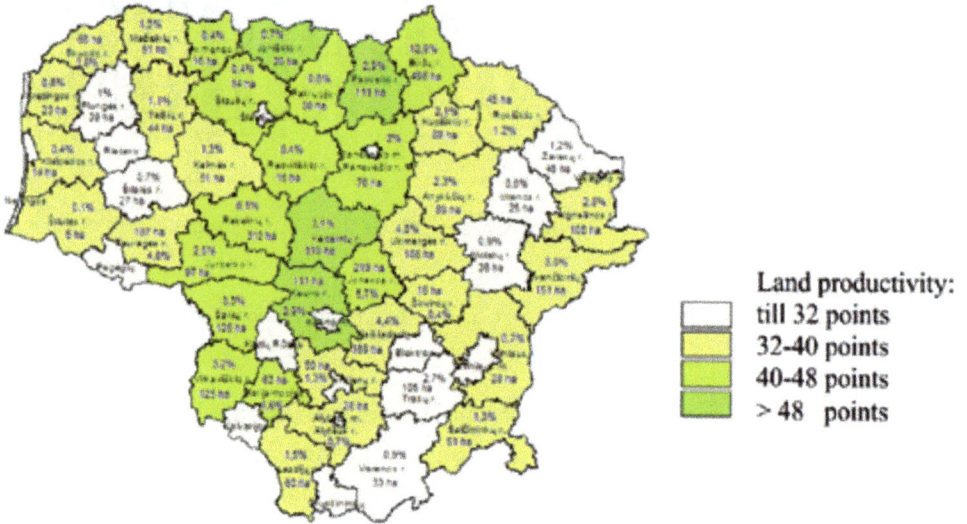

Land productivity:
till 32 points
32-40 points
40-48 points
> 48 points

Conventional wheat area (ha and %) distribution in Lithuania by the
municipalities, 2005

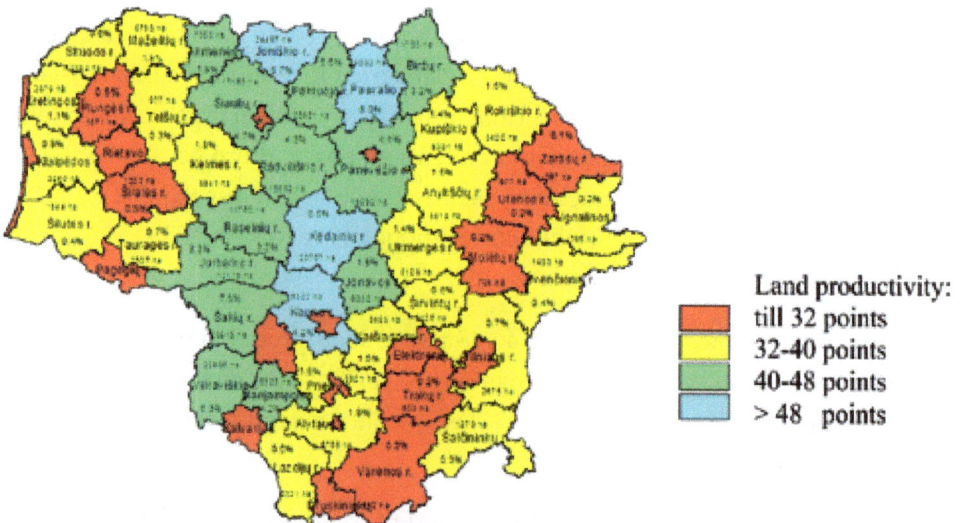

Land productivity:
till 32 points
32-40 points
40-48 points
> 48 points

Figure 17. Organic and conventional wheat area distribution by hectares and percent in Lithuania by the municipalities and land productivity, 2005 [21]

Organic barley area (ha and %) distribution in Lithuania by the municipalities,2005

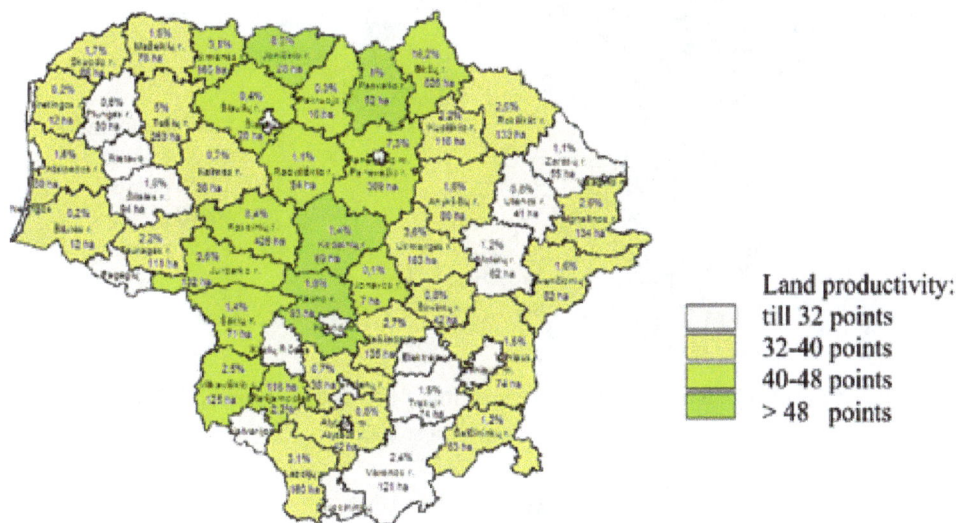

Land productivity:
till 32 points
32-40 points
40-48 points
> 48 points

Conventional barley area (ha and %) distribution in Lithuania by the municipalities,2005

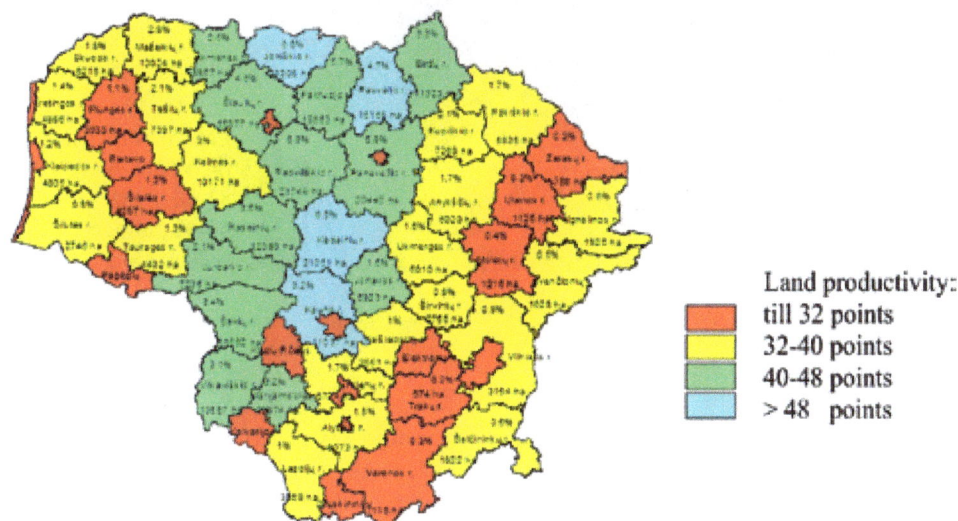

Land productivity:
till 32 points
32-40 points
40-48 points
> 48 points

Figure 18. Organic and conventional barley area distribution by hectares and percent in Lithuania by the municipalities and land productivity, 2005 [21]

Area (ha and %) distribution of organic perennial grasses in Lithuania by the
municipalities, 2005

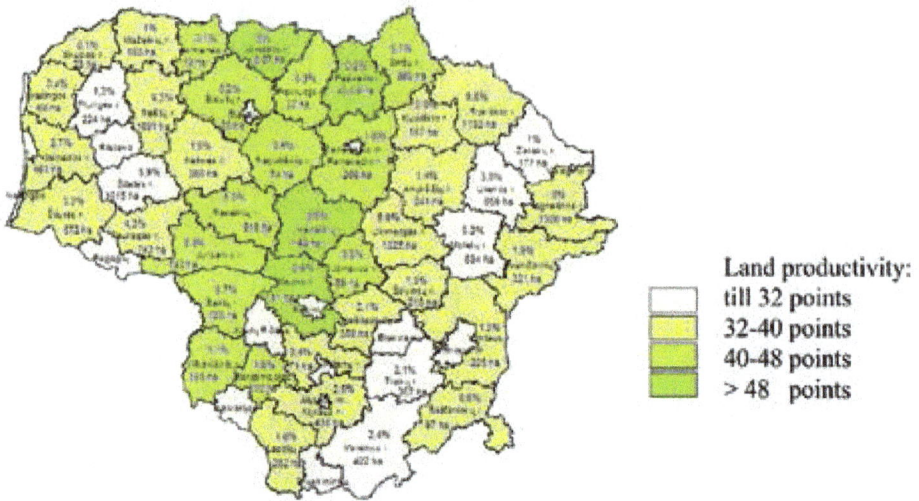

Land productivity:
till 32 points
32-40 points
40-48 points
> 48   points

Area (ha and %) distribution of conventional perennial grasses in Lithuania by the
municipalities, 2005

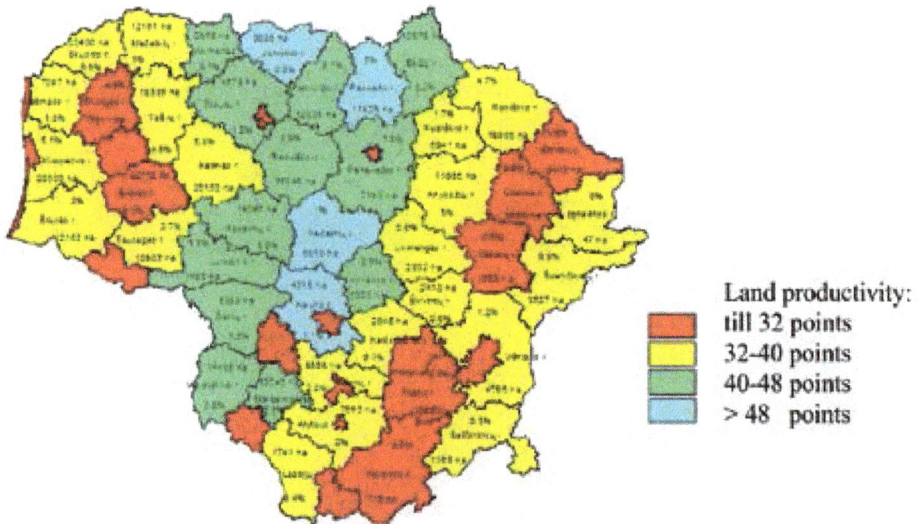

Land productivity:
till 32 points
32-40 points
40-48 points
> 48   points

Figure 19. Area distribution of organic and conventional perennial grasses by hectares and percent in Lithuania by the municipalities and land productivity, 2005 [21]

4.2. Productivity of agricultural crops

The main organic arable crop is cereals (Figure 11-15). Organic cereal yield and productivity in comparison with productivity of conventional grown cereals are presented in table 1. By the grown area spring triticale occupies the least territory while winter rye was the most frequently grown organic cereal in Lithuania in 2011. Productivity of the cereal crop yield [t ha^{-1}] except buckwheat indicated significant decrease of yield productivity in organic agriculture compared it with the conventional one. It is generally known that yield productivity is principally higher in conventional agriculture, because of its industrialization, however, in some crops we received even double size differences. The productivity of organic spring oats was acceptably lower by 23.6% compared it with the conventional oats. Hence, the yield productivity of organic compared it with the conventional winter rye, spring and winter triticale, spring barley, winter and spring wheat was even lower by 61.6%, 75.2%, 84.1%, 102%, 107.5% and 121% respectively. Just average productivity of buckwheat was higher by 20% in organic agriculture compared it with the conventional farming (Table 1).

Index	Cereals, 2011							
	Spring Oat	Spring wheat	Spring triticale	Spring barley	Winter rye	Winter wheat	Winter triticale	Buck-wheat
Organic area, ha	8644	6886	1229	6826	10360	5809	6103	7598
Organic yield, t	14241	10821	1689	10139	12957	9308	8402	9136
Average productivity of organic crop, t ha^{-1}	1.65	1.57	1.37	1.49	1.25	1.60	1.38	1.20
Average productivity of conventional crop, t ha^{-1}	2.04	3.47	2.40	3.01	2.02	3.32	2.54	0.96

Table 1. The productivity of spica cereals grown organically and conventionally in Lithuania 2011. Source: Ekoagros; Statistics Lithuania [22]

The oilseed rape growing is still a problem in organic agriculture. The confirmation of this phenomen is three times less average productivity of organic spring rape and two times less average productivity of organic winter rape compared it with the conventionally grown rapes (Table 2). However, some Lithuanian organic farmers already are producing the organic oilseed rape alimentary oil. The leguminous cereals have high importance in agriculture and especially in organic agriculture because of symbiotic nitrogen fixing bacterium. In this segment of agricultural crops the average yield productivity was higher in conventional than in organic agriculture by 21%-36% and in vetch crop even by 137.5% (Table 2).

Organically grown vegetables need more hand work and request new mashinery and growing technologies, therefore area of organic vegetables are still insignificant (Table 3, Figure 11-15). Difficulties in growing of organic vegetables are reflected on their average productivity. In organic agriculture average productivity of total vegetables is even more than two times lower than in conventional one. The average productivity of organic potatoes in 2011 was 1.5 times lower compared it to the conventional agriculture (Table 3).

Index	Oilseed rape and leguminous cereals, 2011						
	Spring rape	Winter rape	Fodder beans	Peas	Vetch	Lupine	Soya beans
Organic area, ha	1812	90	2260	10594	110	3081	371
Organic yield, t	1198	75	3357	14193	70	2457	264
Average productivity of organic crop, t ha^{-1}	0.66	0.83	1.49	1.34	0.64	0.80	0.71
Average productivity of conventional crop, t ha^{-1}	1.95	1.80	1.81	1.80	1.52	1.09	0.87

Table 2. The productivity of oilseed rape and leguminous cereals grown organically and conventionally in Lithuania 2011. Source: Ekoagros; Statistics Lithuania [22]

Index	Vegetables, 2011							
	Total vegetables	White cabbages	Red beets	Garlics	Pumpkins	Carrots	Onions	Potatoes
Organic area, ha	64.46	4.30	3.14	0.94	1.57	7.49	1.52	346.29
Organic yield, t	669.24	66.82	42.73	3.00	29.33	176.28	17.54	3684.6
Average productivity of organic crop, t ha^{-1}	10.38	15.51	13.61	3.19	18.68	23.54	11.54	10.64
Average productivity of conventional crop, t ha^{-1}	21.33	-	-	-	-	-	-	15.57

Table 3. The productivity of vegetables grown organically and conventionally in Lithuania 2011. Note.-data not available. Source: Ekoagros; Statistics Lithuania [22]

4.3. Organic seed growing

Organic seed material take the special place in organic farming. By the official regulation seed material for organic agriculture must be certified organically and should come from the special farms of organic seed material growers. The exceptions are allowed only if organic seeds are not available in the market by the objective conditions. Anyway, the use of any synthetical chemical stains for seed staining is strongly forbidden. Supply of organic seeds to Lithuanian market from the local certified organic seed growers during time period of 2006-2013 is presented in figure 20. The local specialised organic seed growers still are not able to cover demand of organic seeds. Therefore, near the local organic seed production (Figure 20) organic seeds from abroad are continuously imported (the import data are not available). Organic farmers are obligated every five years to renew the seed material from the special organic seed growing farms or enterprises. Normally, during five year period after seed material renewing, farmers use part of their own crop yield as a seed material for the next season.

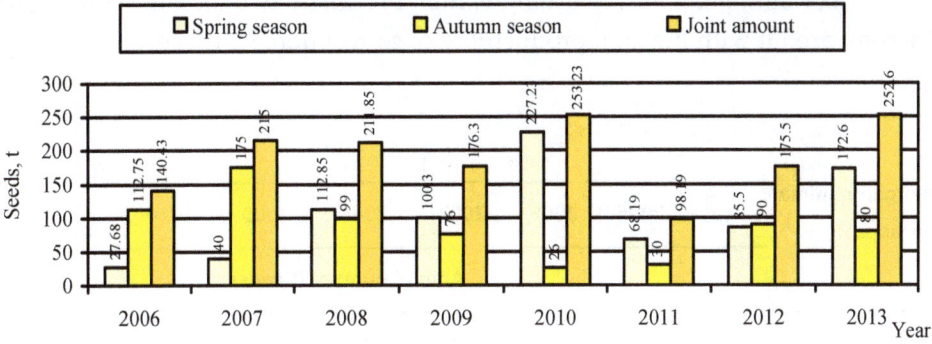

Figure 20. Organic seed amount in tons in Lithuanian market in spring and autumn seasons and joint amount during 2006-2013 [23], Source: Ekoagros.

5. Animal husbandry

The highest farm number where was certified organic livestock was reached in 2012 (Figure 21). It covered 888 farms and compounded 35% from the total certified organic farm number in Lithuania. Similar to EU (Figure 22) cattle and sheep are the most popular species from organic livestock in Lithuania (Table 4). Total certified organic livestock number in Lithuania constantly grew from 2004 to 2012 (except rabbits). The increase of organic animal number (heads) mostly was influenced with coming new certified organic livestock farms to market and only a part of observed increase was induced with development of early organically certified farms.

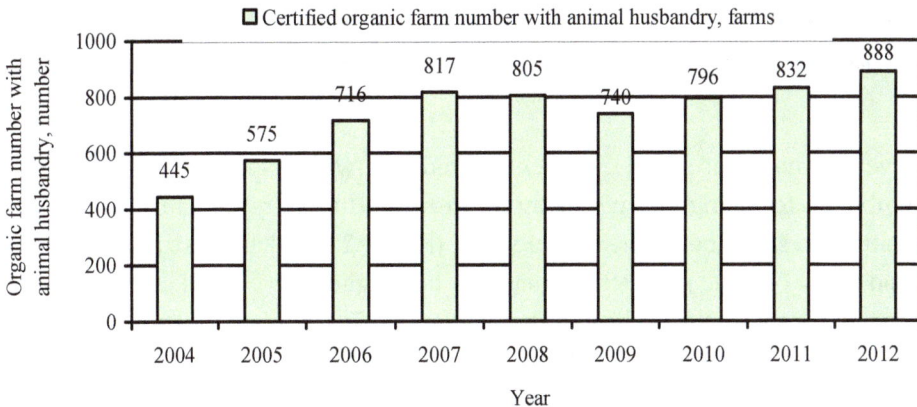

Figure 21. Dynamics number of certified organic farms with livestock during 2004-2012 [18], Source: Ekoagros.

Evaluating cattle's density in 2012 there were established 0.25 livestock units per hectare in conventional agriculture and 0.2 livestock unit per hectare in organic agriculture. The organic milk productivity made 5.52 tons of milk per cow in 2011 while conventional milk productivity made 4.90 tons of milk per cow. Accordingly, milk productivity of organic cows was higher by 12.6% compared it with the milk productivity of conventional farm cows.

Certified organic animal groups	Year								
	2004	2005	2006	2007	2008	2009	2010	2011	2012
	Animal number (heads)								
Dairy cows	3048	4988	6401	7962	8489	8382	8891	8887	9544
Suckler cows	623	14	1071	1507	1915	2252	2863	3359	4086
Bull breeders	22	29	54	68	86	107	117	134	156
Calves	2923	6255	8662	10427	10605	11262	12752	14082	16798
Horses	190	277	321	386	441	488	364	447	474
Goats	321	549	668	740	869	755	586	640	751
Sheep	3789	5052	8507	10561	10768	13001	13683	14276	18307
Pigs	83	266	200	275	203	279	523	474	453
Rabbits	1093	908	369	239	70	215	185	141	69
Cervidae	-	-	-	-	-	-	-	582	752
Poultry	890	1182	344	1121	1100	1510	2709	4406	4103

Table 4. Certified organic livestock (animal number, heads) and its dynamics during 2004-2012 in Lithuania. Source: Ekoagros.

Organic livestock as a share of all livestock showed that, with respect to cattle, pigs and sheep, some Member States using organic methods were producing remarkably large numbers of animals, cattle and sheep being the most popular (Figure 22) [19]. In Austria 25.7% of the sheep were reared using organic production methods, but organically reared cattle also achieved a noteworthy 17.7% share, the highest in the whole EU-27. Estonia had the highest percentage of the sheep population with 47.3%. Lithuania reached the second highest percentage of the sheep population with 27% in 2008 (Figure 22) while till 2012 it increased by 70% (Table 4). As for organically reared pigs, they accounted for less than 1% in most of the Member States (Figure 22) [19].

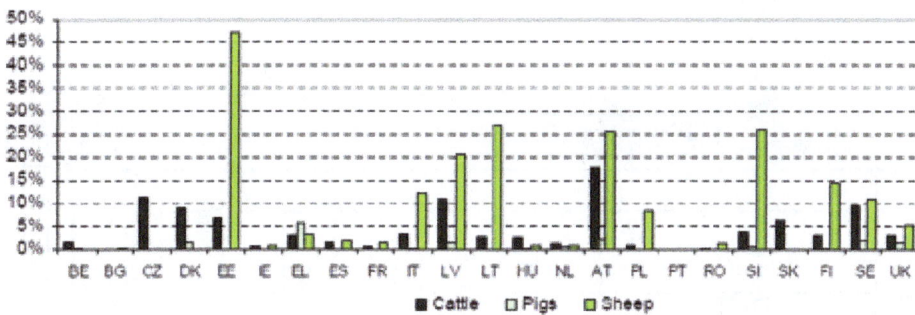

Note. Data on organic livestock: DE, CY, LU, MT no data available; PT all data 2005 [19]
Source: Eurostat (food_in_porg3, apro_mt_lscatl, apro_mt_lspig, apro_mt_lssheep)

Figure 22. Organic livestock (number of heads) out of all livestock in European Union, 2008

6. Processing of organic products

Processing is a very important activity in each sector of economy. It can be as an indicator of viability and development of economy. In organic agriculture firstly it shows enough high quantity of producers that produce at least minimal critical level of primary production. Development of organic processing enterprises and activities by its number dynamics in Lithuania is presented in figure 23. The activities of organic processing enterprises in 2013 were concentrated in grain investment, storage and trading (19 activities), processing (draying, tea production) of medicinal and potherbs (14 activities), wholesale (14 activities), manufacture of grain products (12 activities), milk procurement and processing (7 activities). Some organic processing enterprises entered market in vegetable (6 activities), fruit, berries and mushrooms (6 activities) buying and processing, public catering (4 activities), seed packing and marketing (3 activities), animal slotering and meat products processing (3 activities), fish processing (3 activities) and alimentary oil production (2 activities). One at a time activity of Lithuanian organic processing enterprises was in dumpling, spice, chocolate products, tomato sauce and mayonnaise production and infant nourishment.

Activities within the organic sector include the food chain from production at farm level right through to industrial processing. Imports, exports and other activities, such as wholesale and retail trade, are also included. The production of organic crops and the rearing of organic animals are the main activities in the organic sector at farm level, but the processing of goods is also important. Producers accounted for over 50% of all operators in 2008 in all the Member States and Norway, and even exceeded 70% in most countries. Importers accounted for less than 2% of the total in most of the Member States (Figure 24) [19].

First certificate of organic product importers to Lithuania from the third-countries was issued in 2013. At the end of the year 2013 there were already certified 4 importers organic operators that imported cranberries from Belorussia and the Ukraine, coconut oil from Sri Lanka and etc.

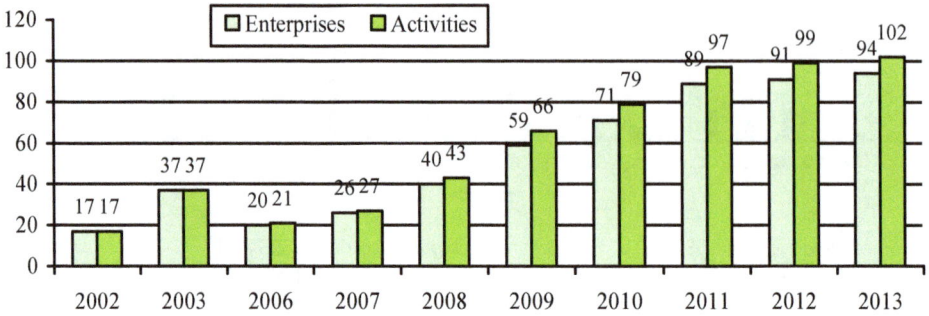

Figure 23. Number change dynamics of organic processing enterprises and activities in Lithuania during 2002-2013. Note. 2013 data of January [18], Source: Ekoagros.

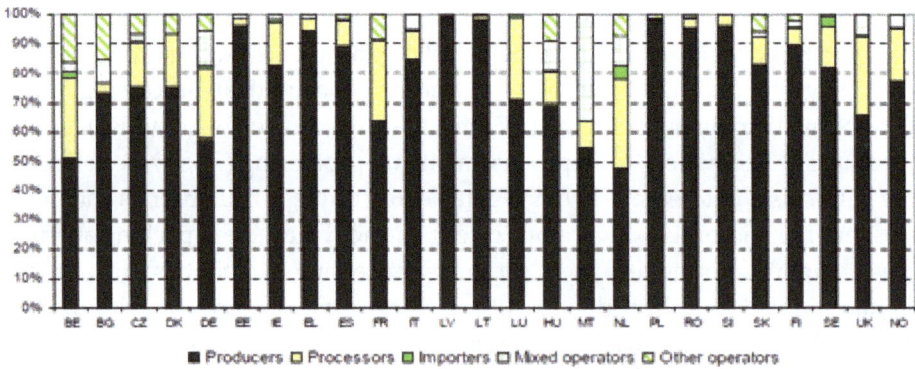

Note. MT data 2006, CY, PT data 2007 [19], Source: Eurostat (food_act2)

Figure 24. Share in percent of different types of operators in total organic operators, 2008

On the basis of the NACE Rev.2 classification, food manufacturing activities can be grouped as follows: processing and preserving of meat and production of meat products, processing and preserving of fruit and vegetables, manufacture of vegetable and animal fats and oils, dairy products, grain mill products and starches, beverages, prepared animal feeds and other food products, including, for example, bakery products, tea, coffee, sugar, chocolate, etc. (Table 5) [19].

Country	Processors by the type of economic activity								
	Meat / meat products	Fruit / vegetables	Vegetables and animal oils / fats	Dairy products	Grain mill products / starches	Other food products	Beverages	Prepared animal feeds	Total
EU-27	2445	4114	2170	1278	1022	6833	2126	519	4633
BE	43	74	9	44	32	247	17	4	470
BG	1	16	7	3	0	10	1	0	38
CZ	72	37	1	31	17	96	69	4	327
DK	53	38	5	36	20	122	29	3	306
DE	-	-	-	-	-	-	-	-	-
EE	2	9	1	2	9	8	0	0	31
IE	33	85	0	5	9	91	3	1	227
EL	72	304	801	74	27	457	214	56	2005
ES	220	520	297	67	49	865	359	38	2415
FR	391	253	50	180	167	372	248	134	1795
IT	308	1277	835	333	353	1871	903	75	2807
CY	-	-	11	-	-	-	3	-	14
LV	3	2	-	5	1	4	-	-	15
LT	3	3	1	5	6	10	0	3	31
LU	4	6	1	2	4	18	1	3	39
HU	-	-	-	-	-	-	-	-	-
MT	0	0	3	0	0	0	1	0	4
NL	138	307	57	101	94	571	60	48	1376
AT	318	145	-	117	71	-	-	56	707
PL	-	-	-	-	-	-	-	-	-
PT12	12	33	23	-	-	2	3	2	75
RO	0	29	2	9	-	24	2	0	66
SI	10	18	3	5	1	20	-	1	58
SK	5	10	2	7	12	35	3	2	76
FI	33	83	4	15	40	52	0	16	243
SE	71	68	18	25	31	116	16	11	356
UK	650	797	39	212	79	1842	194	62	2033
NO	107	64	2	38	28	203	4	10	456

Note. EU-27 estimated; LU data 2004; AT, PT data 2005; MT data 2006; CY data 2007; DE, HU, PL no data available [19], Source: Eurostat (food_act3)

Table 5. Number of organic processors by type of economic activity, NACE Rev. 2, 2008

7. Maintaining of organic agriculture

7.1. Certification

Organic agriculture indicates necessary transformation of farming system therefore it is essential to have a transitional period of 2 years [24]. Transitional period is a period through which on a farm there is introduced crop rotation, fertilization, methods of plant protection and other means of farming corresponding to the regulations of organic agriculture [25]. The certification commission can lengthen or shorten this farm transitional period. Organic agriculture differs from other farming forms as well by the requirement to keep regulations of organic agriculture and its products have origin certification [17, 24]. The whole process of organic product processing, production and all ingredients used in processing is inspected. Certification and marking of organic products initiated its high demand on market. Certification is a procedure by which certification body confirms that product and/or process of processing corresponds to the set of requirements [24, 25]. Production process of organic products control and certified public bodies are validated by the Ministry of Agriculture. In Lithuania organic farms, holdings and enterprises are certified by the public body "Ekoagros" (www.ekoagros.lt). At the moment it is exclusive body for organic agriculture certification in Lithuania with the centre in Kaunas and branches in Utena and Telšiai. However, State Enterprise Lithuanian Agricultural and Food Market Regulation Agency (the Market Regulation Agency <http://www.litfood.lt/Lists/Publications/AllItems.aspx?RootFolder=http%3a%2f%2fwww%2elitfood%2elt%2fLists%2fPublications%2fEnglish%20summary&FolderCTID=0x012000EF8B28BBC9FD604F9F45357A684ABF67>) seek for organic production certification as well. If it succeeds, Lithuania will have two licensed organic certification bodies.

The main standard document of organic agriculture in Lithuania is "Regulations of Organic Agriculture: production, processing, realization and marking" [26] that is continuously improved and renewed [17, 25]. The certification body controls keeping of organic agriculture regulations. Certification body inspects declarant, seeking for organic certificate, performs expertise of inspection results and initiates decission for issue of organic certificate. Certificate is a document issued according to the regulations of organic agriculture and evidenced that product or processing process is in accordance with the requirements of organic agriculture regulations. Certificate gives right to mark products declarated in certificate as organic. The list of certified declarants (farms, holdings, enterprises, etc.) is announced in public. Certificate is valid for one year [25]. All declarants intended to certify production as organic each year till 15 June deliver application for certification and support. After 15 June applications for certification are not admitted. Submitted application data can be corrected till 12 of July. Certification body performs inspection of organic farms till 15 of October each year.

7.2. Support for organic agriculture

Owners of Lithuanian organic farms can receive financial support according to the one of "Agrarian environmental protection payoff" implements programme, i.e. "Organic agriculture". The task of the programme is to support organic agriculture as a system that secures production of high quality products with good perspectives on a market. Support for organic agriculture can be delivered only for the organically certified and declarated agriculture area. Therefore, all owners of the organic farms must be registered and should contain certificate obtained from the organic agriculture certification body "Ekoagros". The exact level of support payoff varies for each organic farmer and is calculated individually depending on growing crops (Table 6). Separate organic farm can obtain maximal support of 400 thousand litas. During 2007-2013 farmers participating in "Rural development programme" implement "Agrarian environmental protection payoff" programme "Organic agriculture" submitted 12 859 applications with requests of 505.5 million litas payoff. According to "Organic agriculture" programme during 2013 there were submited 2566 applications with request of 118 million litas payoff.

Crop	Payout, ** Lt ha⁻¹
Cereals	742
Vegetables and potatoes	1519
Medicinal herbs	1688
Small-fruit plants and orchards	1781
Perennial grasses and meadows*	438

Note. *-subsidies are paid only if there is certified organic livestock on farm; **-1 Lt=0.290 €; 1 €=3.4528 Lt.

Table 6. Support for organic agriculture according to the type of crop (Lt ha⁻¹)

To enhance marketable organic farming is foreseen compulsory realization of organic production (Table 7). Realization of organic production should be validated by actual documents. Only then financial support is delivered for organic farm in form of subsidies. The request to present documents of production realization is not applicable for farms keeping livestock and declarating just pastures and meadows, annual and perennial grasses. As well documents for production realization are not requested for cereals on farm applying proportion 1 LSU (Livestock Standard Unit) per 3 hectares. Subsidies are paid in two stages: 50% of subsidy are paid after evaluation of applications and rest 50% of subsidy are paid after delivered documents of organic production realization. For the new orchards and small-fruit perennial plantations in the first year there are paid 100% of subsidy.

Owners of Lithuanian organic farms can apply for subsidy if:

• they are applicable subjects, i.e. farmers, agricultural company or cooperative;

• have registered agricultural holding in the register of agriculture and rural business of Republic of Lithuania;

• joint agricultural and other area applicable for subsidy by the programme "Organic agriculture" are no less than 1 hectare;

• separate field plot for subsidy is no less than 0.1 hectare.

Type of crop	Required sale of production, * Lt ha⁻¹
In suitable for farming areas	
Cereals	350
Vegetables, potatoes, medicinal herbs	1050
Orchards	1600
Small-fruit plants	1000
In less suitable for farming areas	
Cereals	180
Vegetables, potatoes, medicinal herbs	600
Orchards	900
Small-fruit plants	600

Note. *-1 Lt=0.290 €; 1 €=3.4528 Lt. Source: National Paying Agency under the Ministry of Agriculture, http://www.nma.lt/index.php?lang=2

Table 7. Required compulsory realization of organic production for farms participating in programme "Organic agriculture" in suitable and less suitable areas for farming

Owners of Lithuanian organic farms receiving subsidies by the programme "Organic agriculture" contract to:

- participate in the programme "Organic agriculture" and to keep organic requiremets at least five years from the submition of the application;

- to submit application for subsidy and declare crops each year;

- to keep integrated support interconnect requirements;

- to keep the main requirements;

- to return all paid subsidies by "Organic agriculture" programme if implementation of the programme would be suspended before the term (except special circumstances);

- to run organic agriculture on the same agricultural land, i.e. on the same field plots and every year to declare contracted areas;

- within period of contract do not decrease contracted area more than 3% and do not increase it more than 2 ha;

- to run accountancy according to the regulations of law;

- to have and implement fertilization plans if fertilized area by manure or slurry exceeds 50 ha on farm during artificial year;

- to fill-in journal of organic agriculture production if fertilized area by manure or slurry do not exceed 50 ha on farm during artificial year, plant protection means for non-professional use and mineral fertilizers are used;

- to fill-in journal of applied means and products of plant protection if there are used plant protection means for non-professional use on farm;

- to keep regulations and requirements for organic agriculture foreseen at the EU Council Regulation (EC) No 834/2007 the whole contracted period;

- to sell a part or organic production;

- to keep regulations of Lithuanian organic agriculture.

7.3. Education in organic agriculture

Education, vocational training and, more generally, lifelong learning play a vital role in the economic and social strategies of the European Union [15]. Education is very important in all areas of life. Agricultural professionals play an important role in helping to create and develop innovations. They also inform and educate farmers (and the public) about innovations through teaching or extension work [7].

Inceptive organic farming farmers must keep regulations of Lithuanian organic agriculture "Regulations of organic agriculture" [26] and must take part in the course of educational programme "Backgrounds of organic agriculture (for beginners)" [27]. The course completion certificate must be delivered for the certification body before the organic certificate issue day (i.e. till the 15th June of current year) no later than till the day of farm certification. The educational courses for farmers are administrated by The Centre for LEADER Programme and Agricultural Training Methodology. The tasks of educational course programme "Backgrounds of organic agriculture" is to convey for farmers scientifically and practically validated recommendation of organic agriculture, to present backgrounds of organic production and to acquaint with the main requirements of organic agriculture. The earlier received adequate education to the farmers course programme "Backgrounds of organic agriculture" can be recognized by the committee formed by the The Centre for LEADER Programme and Agricultural Training Methodology under the Ministry of Agriculture.

The State supported trainings (Table 8) as „New technologies in farms of organic production" [28], „Organic horticulture" [29], „Weed control system in organic agriculture" [30] and etc. are also popular between organic farmers. The organic training courses for farmers are organized by the demand according to the educational course programmes confirmed by the Ministry of Agriculture (Table 8).

Educational programme	Year					
	2008	2009	2010	2011	2012	2013
	Number of training courses					
Backgrounds of organic agriculture (for beginners) 24 academic hours [27]	17	29	35	38	25	32
Organic agriculture for advanced 16 academic hours) [31]	-	40	39	31	30	19
Organic seed growing (8 academic hours) [32]	-	-	5	5	-	-
Economical evaluation of organic products production marketing on individual and cooperative background (8 academic hours) [33]	-	-	10	8	-	-
Pecularities of organic agriculture by the specialization of production (field day) 4 academic hours [34]	14	16	-	-	-	-
Organic agriculture for advanced (field day) 4 academic hours [35]	-	12	-	-	-	-
Backgrounds of organic beekeeping 10 academic hours [36]	-	-	18	+	-	-
Organic cattle husbandry 10 academic hours [37]	-	-	104	-	-	-
Weed control system in organic agriculture 10 academic hours [30]	-	-	29	13	-	-
Organic horticulture, 10 academic hours [29]	-	-	-	30	-	-
Organic non-traditional animal husbandry and aviculture, 10 academic hours [38]	-	-	-	24	-	-
Swine-breeding in farms of organic production, 10 academic hours [39]	-	-	-	15	-	-

Source: The Centre for LEADER Programme and Agricultural Training Methodology under the Ministry of Agriculture; Chamber of Agriculture of the Republic of Lithuania

Table 8. Organic training courses for farmers in Lithuania during 2008-2013

The average size of the individual course is 14 farmers. During 2008-2013 there were trained more than 5700 farmers interested in organic agriculture. Compulsory trainings "Backgrounds of organic agriculture (for beginners)" for inceptive organic farming were delivered for about 2200 participants.

8. Conclusions

Organic agriculture is a production system that sustains the health of soils, ecosystems and people. Farming is only considered to be organic at the EU level if it complies with Council Regulation.

The area of certified organic agricultural land in the world, EU and Lithuania continuously is tending to increase. EU-28 average made 5.7% of agricultural land as organic in 2012.

Average size of agricultural farm in general is larger in the organic than in conventional sector.

Productivity of crop average yield regularly is lower in organic agriculture compared it with the conventional one.

Cattle and sheep are the most popular species of the organic livestock.

In the manufacture of organic products fruit, vegetables, meat and meat products are dominating.

Abbreviations

EU-28 European Union of 28 Member States

EU-27 European Union of 27 Member States

EU European Union

BE Belgium

BG Bulgaria

CZ Czech Republic

DK Denmark

DE Germany

EE Estonia

IE Ireland

EL Greece

ES Spain

FR France

HR Croatia

IT Italy

CY Cyprus

LV Latvia

LT Lithuania

LU Luxembourg

HU Hungary

MT Malta

NL Netherlands

AT Austria

PL Poland

PT Portugal

RO Romania

SI Slovenia

SK Slovakia

FI Finland

SE Sweden

UK United Kingdom

IS Iceland

LI Liechtenstein

NO Norway

CH Switzerland

NACE Rev.2 classification: Statistical Classification of Economic Activities in the European Community, Rev. 2.

Acknowledgements

We would like to thank Vilma Pilipavičienė for the manuscript English reviewing linguistically.

Author details

Vytautas Pilipavičius[1*] and Alvydas Grigaliūnas[2]

*Address all correspondence to: vytautas.pilipavicius@asu.lt

1 Aleksandras Stulginskis University, Faculty of Agronomy, Institute of Agroecosystems and Soil Sciences, Akademija, Lithuania

2 The Centre for LEADER Programme and Agricultural Training Methodology, under the Ministry of Agriculture of the Republic of Lithuania, Akademija, Lithuania

References

[1] Definition of Organic Agriculture. IFOAM General Assembly in Vignola, Italy in June 2008. Available from: <http://infohub.ifoam.org/en/what-organic/definition-organic-agriculture>

[2] Lamine C. Transition pathways towards a robust ecologization of agriculture and the need for system redesign. Cases from organic farming and IPM. Journal of Rural Studies 2011; 27(2) 209-219.

[3] Aeberhard A., Rist S. Transdisciplinary co-production of knowledge in the development of organic agriculture in Switzerland. Ecological Economics 2009; 68(4) 1171-1181.

[4] Febles-González JM., Tolón-Becerra A., Lastra-Bravo X., Acosta-Valdés X. Cuban agricultural policy in the last 25 years. From conventional to organic agriculture. Land Use Policy 2011; 28(4) 723-735.

[5] Pilipavičius V. Organic agriculture-development and perspective. In: Organic crop production and horticulture. Edited by Motuzas A., Pilipavicius V., Butkus V. Kaunas, 2006. p.6-10. ISBN 995544844X.

[6] Rigby, D., Caceres, D. Organic farming and the sustainability of agricultural systems. Agricultural Systems 2001; 68(1) 21-40.

[7] Wheeler SA. What influences agricultural professionals' views towards organic agriculture? Ecological Economics 2008; 65(1) 145-154.

[8] Wollni M., Andersson C. Spatial patterns of organic agriculture adoption: Evidence from Honduras. Ecological Economics 2014; 97(1) 120-128.

[9] Council Regulation (EC) No 834/2007 of 28 June 2007. on organic production and labelling of organic products and repealing Regulation (EEC) No 2092/91. Official Journal of the European Union. 23 p. Available from: <http://eur-lex.europa.eu/LexUriServ/LexUriServ.do?uri=OJ:L:2007:189:0001:0023:EN:PDF>

[10] Commission Implementing Regulation (EU) No 1030/2013 of 24 October 2013 amending Regulation (EC) No 889/2008 laying down detailed rules for the implementation of Council Regulation (EC) No 834/2007 on organic production and labelling of organic products with regard to organic production, labelling and control. Official Journal of the European Union. P. 1-2. Available from: <http://eur-lex.europa.eu/LexUriServ/LexUriServ.do?uri=OJ:L:2013:283:0015:0016:EN:PDF>

[11] Commission Regulation (EC) No 889/2008. of 5 September 2008. laying down detailed rules for the implementation of Council Regulation (EC) No 834/2007 on organic production and labelling of organic products with regard to organic production, labelling and control. Official Journal of the European Union. 84 p.

Available   from: <http://eur-lex.europa.eu/LexUriServ/LexUriServ.do?uri=OJ:L:
2008:250:0001:0084:EN:PDF>

[12] Commission Implementing Regulation (EU) No 392/2013 of 29 April 2013 amending
Regulation (EC) No 889/2008 as regards the control system for organic production.
Official Journal of the European Union. 10 p. Available from: <http://eur-lex.euro-
pa.eu/LexUriServ/LexUriServ.do?uri=OJ:L:2013:118:0005:0014:en:PDF>

[13] Basalykas A., Bieliukas K., Chomskis V. Lietuvos TSR fizinė geografija / Physical ge-
ography of Lithuania. Vilnius. Vol. 1, 1958. p. 501-504.

[14] Pilipavičius, V.; Romaneckas, K.; Gudauskienė, A. Crop Weediness, Soil Seed Bank
and Yield of Winter Wheat Spelt under Organic Agriculture. Rural development
2013: the 6th international scientific conference, 28-29 November, 2013. Aleksandras
Stulginskis university, Akademija, Kaunas district, Lithuania. Vol. 6, book 2. p.
213-217. Available from: <http://www.asu.lt/rural__development/lt/50217>

[15] Eurostat regional yearbook 2013. European Commission. Eurostat statistical books.
2013. 284. ISSN 1830-9674. doi:10.2785/44451. Cat. No. KS-HA-13-001-EN-C. Available from: <http://ec.europa.eu/eurostat>

[16] Willer H. Organic agriculture in Europe 2009: Production and Market. Research Institute of Organic Agriculture FiBL, Switzerland. BioFach Congress, Nurnrberg, February 18, 2011. Available from: <www.fibl.org>

[17] Pilipavičius V. Alternatyvinės žemdirbystės vystymasis ir perspektyvos Lietuvoje /
Development and perspectives of alternative agriculture in Lithuania. Kaunas,
LŽŪA. 1996. 39.

[18] The results of certification of 2012. 2012 metų sertifikavimo rezultatai. VšĮ Ekoagros,
2013. 24.

[19] Rohner-Thielen E. Agriculture and fisheries. Eurostat. Statistics in focus. ISSN
1977-0316. Catalogue number: KS-SF-10-010-EN-N. European Union, 2010. 12.

[20] Agriculture, forestry and fisheries statistics-2013 edition (2013). European Commission. Luxembourg: Publications Office of the European Union. Theme: Agriculture
and fisheries. Collection: Pocketbooks. 2013. 256 pp. ISBN 978-92-79-33005-6. ISSN
1977-2262. doi:10.2785/45595. Cat. No: KS-FK-13-001-EN-C. Available from: <http://
ec.europa.eu/eurostat>

[21] Kriščiukaitienė I. Ekologinio ūkio ekonominė perspektyva / Economical perspective
of organic farm. LAEI, Ūkių ir įmonių ekonomikos skyrius. 2006. 35.

[22] Statistics Lithuania (2014). Statistikos departamentas prie LRV (2014). M5010302: Agricultural crops in country / Žemės ūkio augalai šalyje. Available from: <http://
db1.stat.gov.lt/statbank/default.asp?w=1280>

[23] Supply of organic seeds to market 2006-2013 / Ekologiškų sėklų pasiūla rinkai 2006-2013 m. Ekoagros. 2013. 4.

[24] Pilipavičius V. Certification of Ecological Products. Rural Development: Contents, Models and Policies in the E.U. Towards the 21st Century. Edited by Vacca M. Seminar in Italy, Perugia, 18-19 June, 1999. 172-177.

[25] Pilipavičius V. Žemdirbystės sistemos ir jų raida / Farming systems and their development. In: Agronomijos pagrindai. Edited by Romaneckas K. Akademija. Aleksandro Stulginskio universitetas. 2011. 131-148.

[26] Regulations of Organic Agriculture (2013) / Ekologinio žemės ūkio taisyklės (2013). Valstybės žinios. LR žemės ūkio ministro 2000 gruodžio 28 d. įsakymas Nr. 375 (LR žemės ūkio ministro įsakymo Nr.222, Nr.190, Nr.3D–253, 3D–324, 254, 284, 3D–436, 3D–82, 3D–156, 3D–368, 3D–391, 3D–61, 3D-199, 3D–11, 3D–496, 3D–695, 3D–717, 3D–876, 3D-348, 3D-11 (2013 05 16 actual edition).

[27] Backgrounds of organic agriculture (for beginners) 24 academic hours / Ekologinio ūkininkavimo pagrindai (pradedantiems) 24 akad. val. Registration code No. 396185007. Ministry of Agriculture of the Republic of Lithuania. Renewed at 2011-07-22. 2011. 2. Available from: <http://www.zmmc.lt/lt/zemdirbiu-mokymai/mokymo-programos/174-1-mokymo-kryptis.html>

[28] New technologies in farms of organic production 16 academic hours / Naujausios technologijos ekologinės gamybos ūkiuose 16 akad. val. Registration code No. 296185008. Ministry of Agriculture of the Republic of Lithuania. Confirmed at 2008-07-02. 2008.

[29] Organic horticulture 10 academic hours / Ekologinė sodininkystė ir uogininkystė 10 akad. val. Registration code No. 296162239. Ministry of Agriculture of the Republic of Lithuania. Confirmed at 2010-12-09. 2010.

[30] Weed control system in organic agriculture 10 academic hours / Piktžolių kontrolės sistema ekologinėje žemdirbystėje 10 akad. val. Programm prepared by Pilipavičius V. Registration code No. 296162136. Ministry of Agriculture of the Republic of Lithuania. Confirmed at 2009-09-11. 2009.

[31] Organic agriculture for advanced 16 academic hours / Ekologinis ūkininkavimas pažengusiems 16 kad. val. Registration code No. 396185007. Ministry of Agriculture of the Republic of Lithuania. Renewed at 2011-07-22. 2011. 2.

[32] Organic seed growing 8 academic hours / Ekologinė sėklininkystė 8 akad. val. Registration code No. 396185010. Ministry of Agriculture of the Republic of Lithuania. Confirmed at 2010-01-25. 2010. 1. Available from: <http://www.zmmc.lt/lt/zemdirbiu-mokymai/mokymo-programos/174-1-mokymo-kryptis.html>

[33] Economical evaluation of organic products production marketing on individual and cooperative background 8 academic hours / Ekologiškų produktų gamybos ir pardavimo individualiai ir kooperacijos pagrindais ekonominis vertinimas 8 akad. val.

Registration code No. 396134502. Ministry of Agriculture of the Republic of Lithuania. Confirmed at 2010-01-25. 2010.

[34] Pecularities of organic agriculture by the specialization of production (field day) 4 academic hours / Ekologinio ūkininkavimo ypatumai pagal gamybos specializaciją (lauko dienos) 4 akad. val. Ministry of Agriculture of the Republic of Lithuania. 2008.

[35] Organic agriculture for advanced (field day) 4 academic hours / Ekologinis ūkininkavimas pažengusiems (lauko dienos) 4 akad. val. Ministry of Agriculture of the Republic of Lithuania. 2008.

[36] Backgrounds of organic beekeeping 10 academic hours / Ekologinės bitininkystės pagrindai 10 akad. val. Registration code No. 296162134. Ministry of Agriculture of the Republic of Lithuania. Confirmed at 2009-09-11. 2009.

[37] Organic cattle husbandry 10 academic hours / Ekologinė galvijininkystė 10 akad. val. Registration code No. 296162135. Ministry of Agriculture of the Republic of Lithuania. Confirmed at 2009-09-11. 2009.

[38] Organic non-traditional animal husbandry and aviculture 10 academic hours / Ekologinė netradicinė gyvulininkystė ir paukštininkystė 10 akad. val. Registration code 296162158. Ministry of Agriculture of the Republic of Lithuania. 2008.

[39] Swine-breeding in farms of organic production 10 academic hours / Kiaulių auginimas ekologinės gamybos ūkiuose 10 akad. val. Registration code 296162159. Ministry of Agriculture of the Republic of Lithuania. 2008.

6

Tomato Fruit Quality from Organic and Conventional Production

Ilić S. Zoran, Kapoulas Nikolaos and Šunić Ljubomir

1. Introduction

The tomato (*Solanum lycopersicum* L.) is one of the world's most important vegetables, with an estimated total production of about 159.347 million tonnes in 2011 (FAOSTAT 2011). It is the second most widely consumed vegetable after the potato [1]. Tomatoes are important not only because of the large amount consumed, but also because of their high health and nutritional contributions to humans. The tomato processing industry has made tremendous advances, developing many forms of tomato-based foods, such as sauces, catsup (ketchup), puree, pastes, soups, juices and juice blends, and canned tomatoes either whole or in diced, sliced, quartered or stewed form [2]. The tomato's attractive color and flavor have made it a dietary staple in many parts of the world. Nutritional considerations also bring the tomato to the forefront.

In the human diet, it is an important source of micronutrients, certain minerals (notably potassium) and carboxylic acids, including ascorbic, citric, malic, fumaric and oxalic acids [3; 4]. Tomatoes and tomato products are rich in food components that are antioxidant and considered to be a source of carotenoids, in particular lycopene and phenolic compounds [5; 6; 7; 8], but low in fat and calories, as well as being cholesterol-free. Most importantly, tomato consumption has been shown to reduce the risks of cardiovascular disease and certain types of cancer, such as cancers of prostate, lung and stomach [9]. The health promoting benefits of tomatoes and tomato products have been attributed mostly to the significant amount of lycopene contained. The results of various studies suggest that lycopene plays a role in the prevention of different health issues, cardiovascular disorders, digestive tract tumors and in inhibiting prostate carcinoma cell proliferation in humans [10].

As a potent antioxidant, lycopene is presently marketed as a fortified nutritional supplement [2]. Another carotenoid, β-carotene, a precursor of vitamin A, is also abundant in tomato. The carotenoid content of tomatoes is affected by cultural practices on one side – genotype and agronomic technique [11; 12] on the other side. The levels of carotenoids and phenolics are very variable and may be affected by ripeness, genotype and cultivation methods [13; 14; 15].

Tomato quality is a function of several factors including the choice of cultivar, cultural practices, harvest time and method, storage, and handling procedures. Increased interest in organic tomato production imposed the need to evaluate the quality and nutritional value of organic tomato.

Some studies have shown higher levels of bioactive compounds in organically produced tomato fruits compared to conventional ones, but not all studies have been consistent in this respect [16; 17; 18]. Organic tomatoes achieve higher prices and a guaranteed placement compared to conventional tomatoes [19], because these products are often linked to protecting the environment and to having better quality (taste, storage), and most people believe that they are healthier. Organic system enhanced optimal production level but with higher cost of cultivation (certification procedures, higher cost per unit of fertilizer, phytosanitary treatments applied, more labor etc.), compared with conventional farming.

2. Production methods and fruit quality

Both conventional and organic agricultural practices include combinations of farming practices that vary greatly depending upon region, climate, soils, pests and diseases, and economic factors guiding the particular management practices used on the farm [20].These differences between organic and conventional production are reflected in the fertilizer used (organic-manure; conventional-mineral fertilizer), the number of phytosanitary treatments (larger in organic system), and the pesticide types applied (preventive in the organic system and preventive or healing with variable period of effectiveness in the conventional one) [21].

Organic production methods by definition do not guarantee a higher quality product [22]. Research results on the effects of organic and conventional production on fruit quality are sometimes contradictory. In terms of quality, some studies report better taste, higher vitamin C contents and higher levels of other quality related compounds for organically grown products [20; 23], whereas several other studies have found the opposite or no differences in quality characteristics between organically and conventionally grown vegetables [23]. The factors influencing tomato quality are complex and interrelated, and additional studies are necessary to consolidate the knowledge about the real interdependences.

One major problem in comparative studies might be that genuine organic and conventional production systems differ in many factors and that a simple measurement of food composition does not reflect its quality. Other scientists have argued that a valid comparison of nutritional quality would, for example, require that the same cultivars are grown at the same location, in the same soil and with the same amounts of nutrients [24; 25]. However, there is little information on the effect of different forms of cultivation on the antioxidant potential of tomatoes.

3. Materials and methods

Three tomato varieties (Robin F_1, Amati F_1 and Elpida F_1) have been tested in greenhouse production (plastic tunnels 3.5m high, covered with termolux 180 µm) during 2008-2010, located in the Sapes, Northeastern Greece, using two different growing systems: organic and conventional. Greenhouse technology and horticultural practices differ little. The main variations concerned pest control, fertilization and fertility of soil, which was of much better quality in the organic production. In conventional cultivation mineral fertilizers and chemical plant protection were applied. The differences between production systems were the fertilizers used (organic: goat manure 3 tonnes/ha; conventional: mineral fertilizer NPK (12:12:17), nitrophos blue special+2MgO+8S+Trace elements – 400 kg/ha), the number of phytosanitary (solarization) treatments (larger in organic system), the pesticide types applied (preventive in the organic systems and preventive or healing with variable period of effectiveness in the conventional one). It was an early-medium production; planting was done between 15 April and 20 April at a density of 2.64 plants/m^2.

At the pink stage of ripening determined by visual inspection, samples were collected for quality analyses (colour, firmness, total soluble solids, total sugar, total acidity content of vitamin C, content of carotenoids and lycopene). For sensory evaluation fruits were evaluated by trained descriptive panelists on the day of harvest (red stage). Tomato samples (20 fruits) were collected each year from June till August and were taken from the third to sixth floral branches.

Determination of total soluble solids (TSS) was carried out by a refractometer. The results were reported as °Brix at 20 °C. The titrable acidity (TA) was measured with 5 ml aliquots of juice that were titrated at pH 8.1 with 0.1 N NaOH (required to neutralize the acids of tomatoes in phenolphthalein presence) and the results were expressed as citric acid percentages.

Pigment extraction from tomato fruits, preparation of extracts for analysis and calibration plots of standard components were determined according to a described method [26].

Approximately 0.5 g of freeze-dried sample was weighed into porcelain crucibles that had previously been heated for 3 hr at 550° C and was converted to white ash at this same temperature over 12–18 hr. Each ashed sample was dissolved in 20 mL of 3 M HCl, and K, Ca, Na, Mg, Fe, Zn, Mn and Cu levels were determined by atomic absorption spectrophotometry.

Besides, a taste index and the maturity were calculated using the equation proposed by X et al. [27] and Y and co-workers [28] starting from the Brix degree and acidity values which were determined in a previous paper [29].

$$\text{Taste index} = \frac{\text{Brix degree}}{20 \times \text{Acidity}} + \text{Acidity} \qquad \text{Maturity} = \frac{\text{Brix degree}}{\text{Acidity}}$$

4. Phytochemicals

The levels of some phenolic compounds are known to be higher in organic fruit. Plants create phenolic compounds for many reasons, but a major reason is to make plant tissues less attractive to herbivores, insects and other predators. Accordingly, it is important to sort out if higher levels of phenolic compounds affect the taste of organic fruits and vegetables when compared to conventionally grown produce [30].

The organic growing system affects tomato quality parameters such as nutritional value and phenolic compound content. The effect of variety, season, harvest time, maturity, as well as environmental factors such as light, water and nutrient supply on the antioxidant content of tomatoes are reviewed by Dumas et al. [31].

Vitamin C of tomato fruits accounts for up to 40% of the recommended dietary allowance for human beings. Farm management skills combined with site-specific effects contribute to high vitamin C levels, and the choice of variety significantly influences the content of ascorbic acid [32]. The variation in vitamin C content in tomatoes depends mainly on environmental conditions. Exposure to light is a favorable factor for ascorbic acid accumulation [31; 33]. Therefore, it is important to compare organic and conventional foods that are planted and harvested during the same season of the year and that originate from regions with similar incidence of solar radiation.

Ascorbic acid content in organically fertilized tomatoes ranges between 29% and 31% [23, 34], which is higher than the results obtained from tomatoes that were fertilized with mineral solutions. Similarly, ascorbic acid content in tomatoes cultivated with an organic substrate was higher than hydroponically cultivated tomatoes [35]. Many citations from literature confirm that tomatoes coming from organic cultivation procedures present higher vitamin C content than fruits from conventional cultivation [36; 37]. It was also found that fertilizer that was rich in soluble nitrogen (N) could cause a decrease in the ascorbic acid content, probably for indirect reasons, since the nitrogen supply increased the plants' leaf density, which promoted shading over the fruits.

Amati		Robin		Elpida	
Organic	Conventional	Organic	Conventional	Organic	Conventional
11.73	13.8	12.6	13.87	14.33	13.7
LSD 5% 5.69 3.15 6.13					
LSD 1% 13.12 7.26 14.15					

Table 1. Vitamin C content (mg 100g^{-1}) in tomato at the organic and conventional production system

The results demonstrate consistent differences in vitamin C content between tomato cultivars and method of cultivation. Thus, Elpida' tomato fruit in organic production system contained the highest level of vitamin C. Irrespective of the cultivation method used, 'Elpida' on average also contained the highest level of vitamin C (14.3 mg 100g^{-1}) in comparison to the rest of the examined tomato cultivars. The conventionally grown Amati and Robin tomato fruits contained more vitamin C than their organically grown counterparts [38].

5. Lycopene content

The color of the fruits is an important consumer quality parameter. The typical color changes during tomato ripening from green to red are associated with chlorophyll breakdown and the synthesis of carotenoid pigments due to the transformation of chloroplasts to chromoplasts [39]. Pigment synthesis in tomato is closely related to the initiation and progress to ripening and red color of the fruit results from the accumulation of lycopene [40], so that lycopene has been suggested as a good indicator of the level of ripening. Lycopene is considered the predominant carotenoid of tomato fruit (80-90%), followed by β-carotene (5-10%) [41].

The lycopene level of tomato fruit is determined by the genetic potential of the cultivar. Most commonly, lycopene levels range within 4.9 and 12.7 mg 100g^{-1} [42] or between 3.5 and 6.9 mg100g^{-1}fresh weight (f.w.) [43]. Lycopene content ranged from 4.3 to 116.7 mg kg^{-1} on a fresh weight basis, with cherry tomato types having the highest lycopene content [11].The distribution of lycopene in the tomato fruit is not uniform. The skin of the tomato fruit contains high levels of lycopene, comprising an average of 37% of the total fruit lycopene content [45], or 3- to 6-fold higher than in whole tomato pulp [44]. About 12 mg of lycopene per 100 g fresh weight was found in tomato skin, while the whole tomato fruit contained only 3.4 mg 100g^{-1} fresh weight [47]. The outer pericarp constitutes the largest amount of total carotenoids and lycopene, while the locule contains a high proportion of carotene [46].

The lycopene content of tomato fruit also varies due to growing and environmental conditions, mainly temperature and light. In general, field-grown tomatoes have higher levels of lycopene than greenhouse grown fruit [13].The lycopene content determined in 39 tomato genotype varieties ranged from 0.6 to 6.4 mg/100 g and 0.4 to 11.7 mg/100 g for greenhouse and field-grown tomatoes, respectively [11]. Similarly, different cultivar varieties have been shown to possess varied lycopene concentrations [13; 48; 49; 45]. Fruits from the indeterminate tomato cultivar Daniela grown in the greenhouse had a higher lycopene content than field grown fruit [50]. Lycopene content also changes significantly during maturation and accumulates mainly in the deep red stage [51].

Tomatoes grown organically contained substantial amounts of lycopene when ripened to firm red or soft red stages. About half of the total lycopene found in soft red tomatoes was present in pink tomatoes and 70 percent in light-red fruit. Fruit picked at unripe stages (breaker through light red) gained as much or more lycopene as those picked at the firm or soft red stages. Results indicate that fruit could be harvested well before full visible red color without loss of lycopene [52].

Tomatoes grown by the conventional or organic agricultural practices did not show any significant difference in the carotenoid content [23].Thus, the absence of any difference between the organic and conventional tomatoes could be due to the control over the ripening, transportation and storage conditions [54].

The results showed that the lycopene content in organic tomatoes was higher than in conventional tomatoes. The average content of this pigment in the organic fruit was 2.92 mg 100^{-1} g f.w., while for conventional tomatoes it was 2.84 mg 100^{-1}g f.w. (Fig. 1).

Different tomato cultivars produce different lycopene levels. 'Elpida' in organic production contained more lycopene in fruit than the other two cultivars (3.75 mg 100^{-1}g f.w.). Differences in sunlight and temperature between the years might be a cause for the contradictory observations.

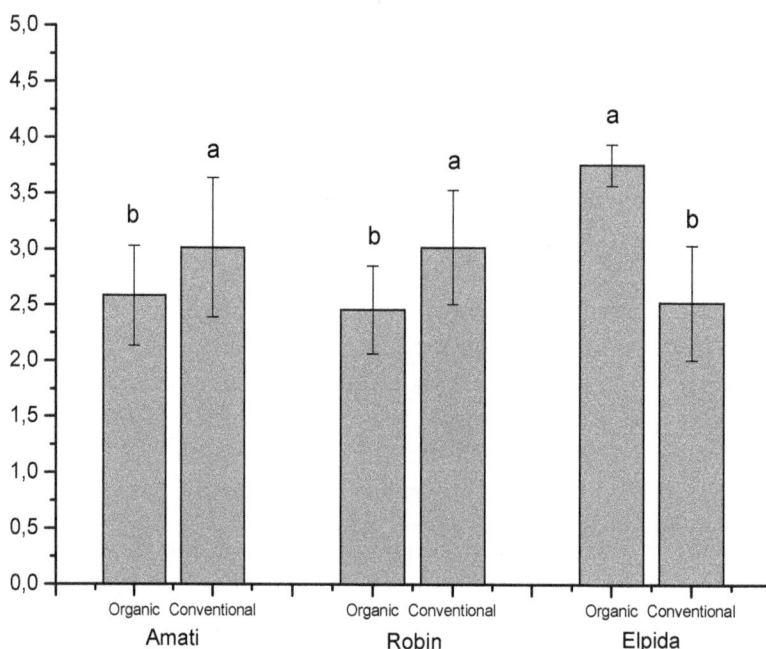

Figure 1. Lycopene content (mg $100g^{-1}$) in organic and conventional tomato cultivars

Tomatoes from organic cultivation contained more carotenoids compared to conventional cultivation. The cultivar 'Amati' contained the lowest level of carotenoids in fruit in both cultivation systems. These differences were statistically significant (p=005). Organically grown 'Robin' produced the highest level of carotenoids in fruit (4.03 mg $100g^{-1}$) comparing to the other two cultivars (Fig. 2).

Studies on carotene and lycopene contents in organic tomatoes, have reported different results including both higher levels [52] and lower levels [53] when compared with conventional methods. No consistent effect of the farming system on the content of bioactive antioxidant compounds [32; 19] was also reported.

Differences between organic and conventional tomatoes can be explained by the fertilizer used in both cases. 'Organic farming doesn't use nitrogenous fertilizers; as a result, plants respond

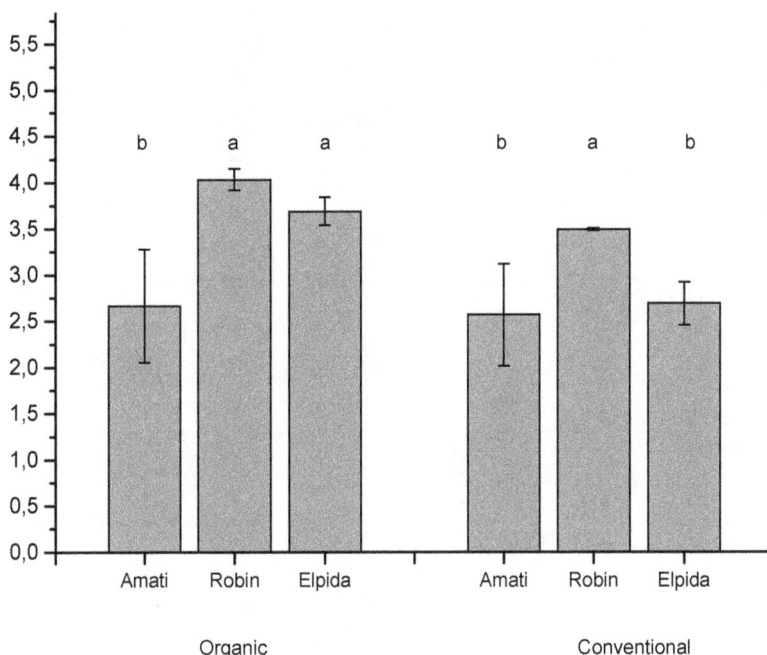

Figure 2. Carotenoids content (mg 100g⁻¹) in organic and conventional tomato cultivars

by activating their own defense mechanisms, increasing the levels of all antioxidants. The more stress plants suffer, the more polyphenols they produce,' these authors point out [55]. Tomato fruits from organic farming experienced stressing conditions that resulted in oxidative stress and the accumulation of higher concentrations of soluble solids as sugars and other compounds contributing to fruit nutritional quality such as vitamin C and phenolic compounds [56].

Flavonoid content in tomatoes seems to be related to available N [34]. Plants with limited N accumulate more flavonoids than those that are well-supplied. If differences in flavonoid content reflect fundamental differences in the behavior of soil N between conventional and organic systems, then the N available to tomatoes late in the season may have declined in organic plots in recent years in response to the cumulative effects of a decrease in compost application rates [57].

Interestingly, yellow flavonoids and anthocyanins did not follow the pattern of total phenolics (Table 2). For instance, the concentration of yellow flavonoids was 70% higher in organic fruits when compared to fruits from conventional growing system, but only at the harvesting stage, which is consistent with similar observations previously [57]. The concentration in anthocyanins was lower in the fruits from organic farming at all three stages of fruit development [56]. These discrepancies indicate that organic farming had the effect of modifying the levels of transcripts or the activities of enzymes controlling intermediary steps of the biosynthetic pathway of phenolic compounds. In spite of the changes in antioxidants, the total antioxidant activity was not significantly different among the organic and conventional tomatoes (Table 2).

Stage of maturity	Organic	Conventional
Total phenolics (mg GAE kg⁻¹)		
Immature	308.563.04 Ab	249.165.65 Aa
Mature	508.361.51 Aa	299.862.39 Ba
Ripe	556.565.40 Aa	232.560.62 Ba
Anthocyanins (mg kg⁻¹)		
Immature	5.160.10 Ba	8.060.19 Aa
Mature	2.560.05 Ba	9.060.16 Aa
Ripe	3.660.09 Ba	9.960.11 Aa
Yellow Flavonoids (mg kg⁻¹)		
Immature	27.860.15 Bb	37.460.33 Aa
Mature	26.160.33 Bb	33.360.43 Aab
Ripe	43.760.49 Aa	25.760.33 Bb
Total Vitamin C (mg kg⁻¹)		
Immature	134.16 ±0.20 Ac	89.46±0.05 Bb
Mature	220.56±0.12 Ab	175.36±0.20 Ba
Ripe	264.76±0.40 Aa	170.96±0.16 Ba
Antioxidant Activity (mMTrolox g⁻¹ FW)		
Immature	98.726±38.65 Aa	98.186±30.42Aa
Mature	143.546±44.52 Aa	161.236±6.15 Aa
Ripe	128.346±22.89 Aa	136.286±57.54Aa

(Oliveira et al., 2013)

Table 2. Quality parameters of tomatoes cultivated organically and conventionally

6. Mineral content

Growing method and cultivar had significant influence on K, Ca, Na or Mg contents in tomato fruits. Organic tomatoes achieved significantly greater concentrations of minerals [58]. The main factor influencing tomato micronutrient content was cultivar [59]. We found significantly greater concentrations of P, K, Ca and Mg in organic tomatoes, but in conventionally grown tomato we found greater content of Zn, Fe and Cu [60]. Our results show that the potassium content in organic tomatoes (153.05-164.31 mg 100g⁻¹) is higher than in conventional tomatoes (126.79-142.54 mg 100g⁻¹). Organically grown Elpida produced the highest level of potassium in fruit (164.31 mg100 g⁻¹) comparing to the other two cultivars. Potassium concentrations were similar to those (191.42–236.54 mg 100g⁻¹) [29]. Our results show that the calcium content in organic tomato (8.08-9.00 mg100g⁻¹) is higher than in conventional tomatoes (7.84-8.58 mg 100g⁻¹). Calcium concentrations (15.97– 23.13 mg 100g⁻¹) were higher in the reported literature [59] than those found in our studies. Significantly greater concentrations of Ca and Mg in organic tomatoes also represented [58]. Magnesium concentrations in organic (17.36-22.22 mg 100g⁻¹) and conventional tomato (18.75-19.16 mg 100g⁻¹) were higher than those found (10.30–11.88 mg 100g⁻¹) [59], but similar to those found in a comparable study [29].

The ranges of measured iron concentration in this study were 0.51-0.64 mg 100g⁻¹ in organic and 0.69-0.72 mg 100g⁻¹ in conventional tomato respectively. In another study, the iron concentration was higher: 0.54-1.37 mg 100g⁻¹ [59]. We observed no significant influence of growing method, which in the case of iron is in keeping with earlier findings [62]. On the contrary, significantly greater concentrations of these minerals in organic tomatoes were found in the report by Kelly & Bateman [58].

Copper concentration (0.11-0.13 mg 100g^{-1}) in conventional tomatoes was higher than in organic tomato (0.5-0.7 mg 100g^{-1}). The ranges of measured copper concentrations (0.05-0.11 mg 100g^{-1}) [59] were higher than those reported by [29; 61]. There were no significant differences in zinc concentrations between organic (0.16-0-18 mg 100g^{-1}) and conventional tomatoes (0.18-0.19 mg100g^{-1}). Zinc concentrations (0.14–0.33 mg 100 g^{-1}) were higher [59] than those reported by Hernández-Suárez et al. and Gundersen et al. [29; 61].

	Moisture %	TotalN	P	K	Ca	Mg	B	Mn	Zn	Fe	Cu
		(mg 100g^{-1} fresh weight)									
		Conventional production									
Elpida	93.19	191.80	33.74	126.79	7.84	18.75	0.02	0.08	0.19	0.69	0.11
Robin	94.28	214.32	29.18	137.59	8.58	19.16	0.03	0.09	0.18	0.73	0.10
Amati	93.62	223.41	27.10	142.54	8.29	18.81	0.03	0.08	0.19	0.82	0.13
		Organic production									
Elpida	93.27	218.77	43.43	164.31	8.08	22.22	0.03	0.08	0.17	0.64	0.07
Robin	92.86	248.73	46.75	159.17	8.92	22.13	0.03	0.08	0.16	0.59	0.05
Amati	93.57	193.02	45.34	153.05	9.00	17.36	0.03	0.07	0.18	0.51	0.05

Table 3. Mineral contents of conventionally and organically grown tomatoes

We found the growing method to have no influence on zinc content, in agreement with previous observations [62]. On the contrary, significantly greater concentrations of Zn in organic tomatoes were found [58].There were insignificant differences of manganese content between conventional (0.08-0.09 mg 100g^{-1}) and organic tomato (0.07-0.08 mg 100g^{-1}). Manganese concentrations (0.05–0.13 mg 100g^{-1}) found by [59], were similar to those reported by [61] and lower than those measured by [29] and were significantly influenced by both cultivar and growing method. Mn levels seem unaffected by the growing method [62]. We found the growing method to have no influence on zinc content (Table 3) like [62]. On the contrary, significantly greater concentrations of Zn in conventional tomatoes were found [58]. On the other hand, in the present study, one possible hypothesis that may explain the insignificant differences in the majority of the minerals could be that the tomato plants of the two cultivation methods managed to have similar soil conditions and irrigation. Previous studies support such a claim. Significant differences in the concentration of Na, Ca, Mg and Zn in tomatoes grown in two different production regions of the island of Tenerife (Spain) have been reported [29]. Some mineral contents in the tomato fruit must be influenced by the region of production, which is mainly influenced by the mineral contents of the cropping soils and of the water for irrigation [29].

7. Index of maturity and taste index

The organic acid in a tomato fruit consist of mainly citric and malic acid with a range of 0.3 to 0.6%. Conventional tomatoes contained more organic acids in comparison to those cultivated by organic methods, in all periods of analysis, being approximately about 0.48% [21].

	Organic production		Conventional production	
	TA (%)	TSS (Brix$^\circ$)	TA (%)	TSS (Brix$^\circ$)
Amati	0.41± 0.01c	4.83± 0.4b	0.48± 0.02 a	4.95± 0.5a
Robin	0.44± 0.01b	4.76± 0.5b	0.47± 0.02a	4.85± 0.6a
Elpida	0.47± 0.01a	5.08± 0.5a	0.48± 0.01a	4.59± 0.5b

Table 4. Total acidity (TA) and total soluble solid (TSS) content of three tomato cultivars from organic and conventional production system

At the same time, it should be noted that Elpida tomatoes were richer in organic acids in comparison to other examined cultivars, independently from the used cultivation system (Table 4). As with the sugars, the organic acids are crucial to the flavour of the fruits. The average contents Brix degree and acidity were 4.6 and 0.50 g 100^{-1} of citric acid, respectively [64].

The concentration of sugars may vary from 1.66 to 3.99% and 3.05 to 4.65% of the fresh matter, as a function of the cultivar and cultivation conditions, respectively [63]. As with the sugars, the organic acids are crucial to the flavour of the fruits. The average contents Brix degree and acidity were 4.6 and 0.50 g 100g^{-1} of citric acid, respectively [29]. The taste index is calculated by applying the equation using the values of Brix degree and acidity [27].

The Elpida cultivar from organic production system had a mean value (1.1) of the taste index higher (P < 0.05) than those values determined for the organic Robin (0.98) and Amati (1.0) cultivars (Table 5). No significant differences were found between the cultivars in the mean taste index obtained for conventionally cultivated tomatoes. When using these data, the mean values of the taste index in all the tomatoes belonging to all the cultivars considered were higher than 0.85, which indicates that the tomato cultivars analyzed are tasty. If the value of the taste index is lower than 0.7, the tomato is considered as having little taste [27].

Another parameter related with the taste index is the maturity index which is usually a better predictor of an acid's flavour impact than Brix degree or acidity alone. Acidity tends to decrease with the maturity of the fruits while the sugar content increases. Significantly greater maturity index in organic Amati fruit (11.7) and lower maturity index in conventional Elpida fruit (9.6) were found [21]. The maturity index in this study (in all cultivars in both production systems) were higher than those found by maturity index reported by [64] was 9.4 and therefore, it can

	Amati F1		Robin F1		Elpida F1	
	Organic	Conventional	Organic	Conventional	Organic	Conventional
Maturity Index	11.7a	10.3b	10.8b	10.3b	10.8b	9.6b
Taste index	1.00b	1.00b	0.98b	0.98b	1.10a	0.96b

Table 5. Index of maturity and taste index of tomato from organic and conventional production

be deduced that the maturity levels of the analyzed tomatoes were adequate for consumption. This ratio can also be affected by climate, cultivar and horticultural practices [28]. The cultivar is a more influential factor than cultivation methods in the differentiation of the tomato samples according to the chemical characteristics. However, quality is more than this and can be defined as the sum of all characteristics that make a consumer satisfied with the product [65]. Apart from functional and nutritional characteristics, quality can include aspects of production method, environment or ethics, as well as availability of and information about a product [66]. For all nutrients examined, cultivar differences were greater than differences because of cultivation method. This study confirms that the most important variable in the micronutrient content of tomatoes is the cultivar; organically grown tomato is no more nutritious than conventionally grown tomato when soil fertility is well managed [59]. Greenhouse tomato production offers advantages compared to production at the open field with regard to quality assurance principally, because the plants are not exposed directly to the rapid changes of climate conditions. An important role for this purpose is also the cultivar selection by using tomato hybrid varieties with a high yield potential and a good fruit quality.

8. Sensory attributes

During tomato fruit ripening, a series of quantitative and qualitative changes take place, changing tomato flavor and aroma volatile profiles [67; 68]. Regarding aroma, several descriptors are present in tomatoes and volatiles are part of the tomato aroma profile. 3-Methylbutanol is an amino acid related compounds, which has a pungent/earthy aroma [69]. Hexanal is one of the major aldehydes in tomatoes and is considered important for fresh tomato flavour.

Panelists could perceive a difference between conventional and organic tomatoes by smell or taste with high reliability. Organic tomatoes were perceived by some of the panelist to be *softer*, and were preferred because of their taste, flavor, texture and juiciness. Alternatively, conventional tomatoes were described as *'not as ripe'*, *'dry'*, and having *'less aroma'* [19].

Very different patterns of correlation between nonvolatile and volatile components emerged as perceived by panelists, depending on whether the nasal passage was blocked to evaluate taste descriptors. A composite of all data collected over the three seasons revealed the 'sweet' note is positively correlated with soluble solids, total sugars, and sucrose equivalents with partitioning (taste followed by aroma).

Elpida

Odors	Taste	Mouth feeling

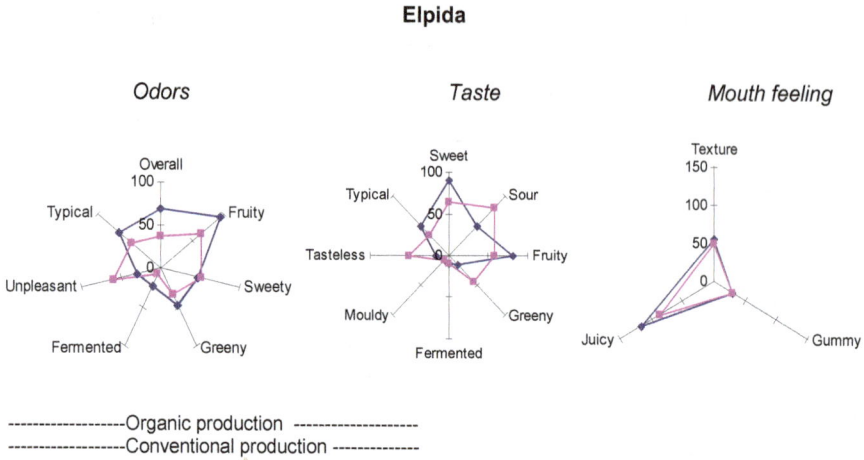

------------------Organic production --------------------
------------------Conventional production --------------

In previous studies, strong positive correlation has been observed between trained panel response of 'sweetness' and reducing sugar and total soluble solids content [70]. Both 'tomato-like' and 'fruity' were positively correlated to acidity and negatively correlated to soluble solids in aroma plus taste trials, but not in the taste followed by aroma trials. A possible explanation in the lack of correlations with many of these descriptors is that there was little difference between these treatments in the lines selected. It is clear that evaluating for taste plus aroma was more sensitive than evaluating for aroma plus taste. It would however be impulsive to conclude that either production system is superior to the other with respect to healthy or nutritional composition [71].

The fruit quality, in terms of taste and nutritional value, did not differ significantly between tomatoes grown in organic or conventional systems. It can take a number of years for soil nutrients to reach optimal levels using organic fertilisers and nutrient availability in the organic systems had probably not been fully established in the three years of the experiments.

However, the type of tomato was more important in determining fruit quality than the type of cropping system: the older variety produced tomatoes with the highest quality index compared with the modern cultivars, implying there is a trade-off between tomato quality and yield [72]. If the aim of organic systems is to produce fruit of superior quality, it is suggested that old cultivars could be used to develop new tomato cultivars adapted for organic cultivation rather than for conventional systems.

9. Heavy metals

Some heavy metals at low doses are essential micronutrients for plants, but in higher doses they may cause metabolic disorders and growth inhibition for most of the plants species [73]. Among the contaminants found in vegetables, heavy metals may reach different levels depending on their content in the soil and the type of fertilization used [73]. For this reason the type of farming techniques can affect the heavy metal content of tomatoes. Both organic (e.g., farmyard manure) and inorganic amendments (e.g., lime, zeolites, and iron oxides) were found to decrease the metal accumulation [74].

The tomato as a fruit vegetable is not characterized by high accumulation of heavy metals. Producers of organic vegetables, do not use mineral fertilizers and practically never use fertilizers produced by industrial waste, which are the most polluted. As a result, one might expect that organic vegetables contain lower amounts of toxic heavy metals. The effect of manure on heavy metal availability is due to the introduction of organic matter to the soil, which may retain Cd in the soil and prevent it from both leaching and from crop uptake [75].

The lead content of tomato fruit, in general, is very low and ranges depending on the hybrid and the methods of production from 0.07 to 0.19 mg 100g^{-1} [19]. No statistical difference in the lead content between organic (0.11 mg 100g^{-1}) and conventional (0.10 mg 100g^{-1}) production of the hybrid Elpida was seen. In the other two hybrids, the lead content was lower in organic production. 'Robin' in organic production achieved lower lead content (0.08 mg kg^{-1}) in comparison with conventional methods (0.10 mg 100g^{-1}). The lead content in 'Amati' is twice lower in organic (0.07 mg kg^{-1}) than in conventional production (0.14 mg 100g^{-1}).

The zinc content in tomato fruits in our studies was just below 20 mg 100g^{-1}. The lower zinc content of the hybrids in organic farming compared to conventional production was not statistically significant. Differences in the content of zinc exist between the individual hybrids. Thus, the lowest zinc content (0.16 mg kg^{-1}) was obtained in 'Robin' in organic production.

	Pb	Zn	Cu	Ni	Cd	Co	Cr
			Organic production				
Elpida	0.11	0.17	0.07	0.01	0.0027	0.0070	0.01
Robin	0.08	0.16	0.05	0.01	0.0027	0.0070	0.01
Amati	0.07	0.18	0.05	0.01	0.0027	0.0070	0.01
			Conventional production				
Elpida	0.14	0.19	0.11	0.02	0.0027	0.0070	0.01
Robin	0.11	0.18	0.10	0.02	0.0027	0.0070	0.01
Amati	0.11	0.19	0.13	0.02	0.0027	0.0070	0.02

Table 6. Heavy metals contents (mg 100g^{-1} f.w.) of conventional and organical tomato

Copper content in organic fruit production is lower, ranging from 0.5 mg 100g^{-1} hybrids Robin and Amati to 0.7 mg 100g^{-1} hybrids Elpida. The copper content in conventional tomato production is twice as high in the hybrids Robin (0.10 mg 100g^{-1}) and Amati (0.13 mg 100g^{-1}) in relation to organic production. The copper content in the hybrid Elpida is 0.11 mg 100g^{-1} (Table 6). In contrast, significantly greater concentrations of Cd (33 μg kg^{-1}) and Pb (37.8 μg kg^{-1}) were found in organic tomatoes, but at the same time a lower Cu content (0.46 mg kg^{-1})

was observed [53]. Systematic fertilization with pig and poultry manure can lead to the accumulation of heavy metals, especially copper.

We found the growing method to have no influence on cadmium (0.0027 mg 100g$^{-1)}$ and cobalt (0.007 mg 100g^{-1}) levels in all cultivars. In the present study, the detected levels of contaminants were found to be markedly lower than the maximum limits allowed by Law: 100 µg kg^{-1} for Pb and 50 µg kg^{-1} for Cd (EU Regulation 1881/2006).

The concentrations of heavy metals in tomato fruit decreased in the order of Zn>Pb>Cu>Cr>Ni>Co>Cd.

10. Nitrate content

Nitrate content of vegetables depends on a number of external and internal factors [76; 77]. From external factors should be mentioned; supply of substrate with nitrate, light, time of day, temperature, season, supply with water, relative humidity, carbon dioxide concentration in the air, supply with biogenic elements, the influence of the accompanying cations, heavy metals, herbicides, chemical properties of the soil, location, time of sowing, time and method of harvest, storage conditions, etc. [78; 79]. Among the internal factors, the most important are the genetic specificity in the accumulation of nitrate (differences between species and differences within genotypes), the distribution of nitrate in certain parts of the plant and the age of the plants..

Nitrate content of various parts of a plant differs [76]. Vegetables that are consumed with their roots, stems and leaves have a high nitrate accumulation (up to 2000 mg kg^{-1}), whereas those with only fruits and melons as consumable parts have a low nitrate accumulation [80]. The tomato belongs to the vegetable plants which accumulate less nitrates than other vegetables (100 to 150 mg kg^{-1}). The effect of climate on nitrate accumulation has been studied [81], and it was found that nitrate content was lower in years that had a high rainfall. In warm and wet years, increased accumulation of nitrate is possible, regardless of whether the nitrogen originates from organic or mineral sources [82]. A comparable study performed in Austria on 17 vegetables found lower nitrate contents (–40% to –86%) in organic vegetables, with spinach being an exception [84]. In Germany, a comparison on carrots showed 61% less nitrates in organic ones [85]. In contrast, two other studies performed on tomato in Israel [83] and carrot in Norway did not show noticeable differences [86].

Nitrogen-rich organic fertilizers can also generate lower nitrate contents, but when mineralization conditions are very favorable they can also lead to high nitrate accumulations [87]. The use of organic fertilization with slowly or moderately available nitrogen (especially composts) is key to explaining the generally observed lower nitrate accumulation in organic vegetables [88].

Differences in nitrate content between cultivars in organic production are present. The lowest nitrate concentration was observed in 'Elpida' (20 mg kg^{-1}) and it was statistically significantly (p<0.05) lower than the nitrate content in the 'Robin' and 'Amati' cultivars. The differences in nitrate content between 'Robin' (27 mg kg⁻¹) and 'Amati' (29 mg kg⁻¹) in organic production were not statistically significant (Fig.3).

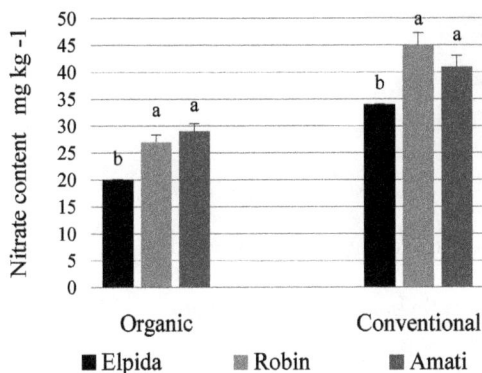

Figure 3. Nitrate content (mg kg⁻¹) in tomato fruit from organic and conventional production

The nitrate content in this study is presented as the average of all cultivars, and it was found to be lower in organic production (29%-41%) compared to conventional production.

In conventional tomato production the nitrate content was lowest in 'Elpida' (34 mg kg⁻¹). The nitrate concentration was significantly ($p < 0.05$) lower than in the other two cultivars. The difference in the nitrate content between the 'Robin' (45 mg kg⁻¹) and 'Amati' (41 mg kg⁻¹) cultivars was not statistically significant.

Rational application of organic manure instead of inorganic nutrients, use of physiologically active substances, proper spray of nitrification inhibitors and molybdenum fertilizers, and growing plants under controlled environmental conditions may all be factors that materially reduce nitrate accumulation in tomatoes.

Selection among the available genotypes/cultivars and breeding of new cultivars that do not accumulate nitrate even under heavy fertilization may also limit human consumption of nitrate through vegetables [89].

11. Conclusion

For all nutrients examined, cultivar differences were greater than differences due to cultivation method. The identification of cultivars with high nutritive value, represent a useful approach to select tomato cultivars with better health-promoting properties.

In general, the significant differences between tomatoes grown in organic or conventional production systems are:

1. organic tomatoes contain more carotenoids

2. organic tomatoes contain more minerals (P, K, Mg, Ca)

3. organic tomatoes contain far less heavy metals (Pb, Zn, Cu, Ni)

4. organic tomatoes contain less nitrates, about 30-40% less

5. organic tomatoes do not contain any pesticide residues

Author details

Ilić S. Zoran[1], Kapoulas Nikolaos[2] and Šunić Ljubomir[1]

*Address all correspondence to: zoran.ilic63@gmail.com

1 Faculty of Agriculture Priština-Lešak, Lešak, Serbia

2 Regional Development Agency of Rodopi, Komotini, Greece

References

[1] Lugasi, A., Bíró, L., Hóvárie, J., Sági, K.V., Brandt, S. & Barna, E. (2003). Lycopene content of foods and lycopene intake in two groups of the Hungarian population. *Nutrition Research*, Vol.23, pp.1035-1044.

[2] Preedy, V. R. & Watson, R. R. (2008). Tomatoes and tomato products: nutritional, medicinal and therapeutic properties. *Science Publishers*. pp. 27-45, USA, ISBN 978-1-57808-534-7.

[3] Caputo, M., Sommella, M. G., Graciani, G., Giordano, I., Fogliano, V., Porta, R. & Mariniello, L. (2004). Antioxidant profiles of corbara small tomatoes during ripening and effects of aqueous extracts on j-774 cell antioxidant enzymes. *Journal of Food Biochemistry*, Vol.28, pp. 1–20.

[4] Hernandez-Suarez, M., Rodrýguez-Rodrýguez, E. M. & Dýaz-Romero C. (2007). Mineral and trace element concentrations in cultivars of tomatoes. *Food Chemistry*, Vol. 104, pp.489-99.

[5] George B., Kaur, C., Khurdiya, D.S. & Kapoor, H.C. (2004). Antioxidants in tomato (*Lycopersium esculentum*) as a function of genotype. *Food Chemistry*, Vol.84, pp.45-51.

[6] Sahlin, E., Savage, G.P. & Lister, C.E. (2004). Investigation of the antioxidant properties of tomatoes after processing. *Journal of Food Composition and Analysis*, Vol.17, pp. 635-647.

[7] Ilahy, R., Hdider, C., Lenucci, M.S, Tlili, I. & Dalessandro, G. (2011). Phytochemical composition and antioxidant activity of high-lycopene tomato (*Solanum lycopersicum* L.) cultivars grown in Southern Italy. *Scientia Horticulturae*, Vol.127, pp.255-261.

[8] Pinela, J., Barros, L., Carvalho, A.M. & Ferreira, I.C.F.R. (2012). Nutritional composition and antioxidant activity of four tomato (*Lycopersicon esculentum* L.) farmer' varieties in Northeastern Portugal homegardens. *Food Chemistry and Toxicology, Vol. 50,* No.(3-4), pp.829-834.

[9] Canene-Adams, K., Campbell, J.K., Zaripheh, S., Jeffery, E.H. & Erdman, J.W. (2005). The tomato as a functional food. *Journal Nutrition,* Vol.135, No,5, pp.1226-1230.

[10] Levy, J. & Sharoni, Y. (2004). The functions of tomato lycopene and its role in human health. *HerbalGram,* Vol. 62, pp.49-56.

[11] Kuti, J.O. & Konuru, H.B. (2005). Effects of genotype and cultivation environment on lycopene content in red-ripe tomatoes. *Journal of Science of Food and Agriculture,* Vol. 85, pp.2021-2026.

[12] Binoy, G., Kaur, C., Khordiya, D.S. & Kapoor, H.C. (2004). Antioxidants in tomato (*Lycopersium esculentum*) as a function of genotype. *Food Chemistry,* Vol. 84, pp.45-51.

[13] Abushita, A.A., Daood, H.G. & Biacs, P.A. (2000). Change in carotenoids and antioxidant vitamins in tomato as a function of varietal and technological factors. *Journal of Agriculture and Food Chemistry,* Vol.48, pp. 2075-2081.

[14] Martínez-Valverde, I., Periago, M.J., Provan, G. & Chesson, A. (2002). Phenolic compounds, lycopene and antioxidant activity in commercial varieties of tomato (*Lycopersicum esculentum*). *Journal of the Science Food and Agriculture,* Vol.82, pp.323-330.

[15] Hallmann, E. (2012). The influence of organic and conventional cultivation systems on the nutritional value and content of bioactive compounds in selected tomato types. *Journal of the Science of Food and Agriculture,* Vol.92, No.14, pp.2840-2848.

[16] Rembiałkowska, E. (2004). The impact of organic agriculture on food quality. *Agricultura,* Vol.1, pp.19-26.

[17] Ordonez-Santos, L. E., Vazquez-Oderiz, M. L. & Romero-Rodrýguez, M. A. (2011). Micronutrient contents in organic and conventional tomatoes (*Solanum lycopersicum* L.). *International Journal of Food Science and Technology,* Vol.46, pp.1561-1568.

[18] Chassy, A.W., Bui, L., Renaud, E.N., Horn, M.V. & Mitchell, A.E. (2006). Three-year comparison of the content of antioxidant microconstituents and several quality characteristics in organic and conventionally managed tomatoes and bell peppers. *Journal of Agricultural and Food Chemistry,* Vol. 54, pp.8244-8252.

[19] Kapoulas, N., Ilić, S. Z., Trajković, R., Milenković, L. & Đurovka, M. (2011). Effect of organic and conventional growing systems on nutritional value and antioxidant activity of tomatoes. *African Journal of Biotechnology,* Vol. 10, No.71, pp.15938-15945.

[20] Mitchell, A E., Yun-Jeong Hong, Koh, E., Barrett, D. M., Bryant, D. E., Denison, R. F. & Kaffka, S. (2007). Ten-Year Comparison of the influence of organic and convention-

al crop management practices on the content of flavonoids in tomatoes. *Journal of Agriculre and Food Chemistry,* Vol.55 No.15, pp.6154-6159.

[21] Kapaulas, N. (2012). The yield and quality of tomato from organic and conventional plastichouse production practices in Northeastern Greece. PhD Thesis, Faculty of Agriculture, Novi Sad

[22] Heeb, A. (2005). Organic or mineral fertilization effects on tomato plant growth and fruit quality. Doctoral thesis, Faculty of Natural Resources and Agricultural Sciences. Swedish University of Agricultural Sciences. Uppsala

[23] Caris-Veyrat, C., Amiot, M. J., Tyssandier, V., Grasselly, D., Buret, M., Mikoljozak, M., Guilland, J. C., Bouteloup-Demange, C. & Borel, P. (2004). Influence of organic versus conventional agricultural practice on the antioxidant microconstituent content of tomatoes and derived purees; consequences on antioxidant plasma status in humans. *Journal of Agricultural and Food Chemistry,* Vol.52, No.6, pp.503-509.

[24] Bourn, D. & Prescott, J. (2002). A comparison of the nutritional value, sensory qualities and food safety of organically and conventionally produced foods, *Critical Reviews in Food Science and Nutrition,* Vol.42, pp.1-34.

[25] Magkos, F., Arvaniti, F. & Zampelas, A. (2006). Organic food: buying more safety or just peace of mind? A critical review of the litterature, *Critical Reviews in Food Science and Nutrition,* Vol.46, pp.23-56.

[26] Cvetkovic, D. & Markovic, D. (2008). UV-induced changes in antioxidant capacities of selected carotenoids toward lecithin in aqueous solution. *Radiation Physics and Chemistry, Vol.* 77, pp. 34-41.

[27] Navez, B., Letard, M., Graselly, D. & Jost, M. (1999). Les criteres de qualite de la tomate. *Infos-Ctifl,* Vol.155, pp.41-47.

[28] Nielsen, S. (2003). *Food Analysis,* 3rd Ed. Kluwer Academic, New York.

[29] Hernández-Suárez, M., Rodríguez-Rodríguez, E. M. & Díaz Romero, C. (2008). Chemical composition of tomato (*Lycopersicon esculentum*) from Tenerife, the Canary Islands. *Food Chemistry,* Vol.106, pp.1046-1056.

[30] Benbrook, C.M. (2005). Elevating antioxidant levels in food through organic farming and food processing. The Organic Center. June 2007.

[31] Dumas, Y., Dadomo, M., Lucca, G., Grolier, P. & Di Luca, G. (2003). Effects of environmental factors and agricultural techniques on antioxidant content of tomatoes. *Journal of the Science of Food and Agriculture,* Vol.83, pp.369-382.

[32] Juroszek, P., Lumpkin, H. M., Yan, R., Ledesma, D. R. & Ma, C. (2009). Fruit quality and bioactive compounds with antioxidant activity of tomatoes grown on-farm: comparison of organic and conventional management systems. *Journal of Agricultural and Food Chemistry, Vol.*57, No.1, pp.188-94.

[33] Martínez-Valverde, I., Periago, M.J., Provan, G. & Chesson, A. (2002). Phenolic compounds, lycopene and antioxidant activity in commercial varieties of tomato (*Lycopersicum esculentum*). Journal of Science Food and Agriculture, Vol.82, pp.323-330.

[34] Toor, R. K., Savage, G. P. & Heeb, A. (2006). Influence of different types of fertilisers on the major antioxidant components of tomatoes. *Journal of Food Composition and Analysis*, Vol.19, pp. 20-27.

[35] Premuzic, Z., Bargiela, M., Garcia, A., Rondina, A. & Lorio, A. (1999). Calcium, iron, potassium, phosphorus, and vitamin C content of organic and hidroponic tomatoes. *Hortscience*, Vol.33, pp.255-257.

[36] Lundegardh, B. & Martensson, A. (2003). Organically produced plant foods – evidence of health benefits. *Acta Agriculturae Scandanavica Section B: Soil & Plant Science*, Vol.53, No.1, pp.3-15.

[37] Borguini, R. G. & Torres, E. A. F. S. (2006). Organic food: nutritional quality and food safety. *Seguranca Alimentare Nutricional*, Vol.13, pp. 64-75.

[38] Kapoulas, N., Ilić, Z. & Đurovka, M. (2011). Tomato quality parameters from organic greenhouse production 46th Croation and 6th Internatoional Symposium on Agriculture. Opatija, Croatia, *Symposium Proceedings*, pp.541-544.

[39] Serrano, M., Zapata, P. J., Guillén, F., Martínez-Romero, D, Castillo, S. & Valero, D. (2008). Post-harvest ripening of tomato. In: Tomatoes and tomato products: nutritional, medicinal and therapeutic properties. Science Publishers. (Editors-Preedy and Watson), pp.67-84, USA

[40] Helyes, L. & Pék, Z. (2006). Tomato fruit quality and content depend on stage of maturity. *HortSciece, Vol.*41, No.6, pp.1400-1401.

[41] Lenucci, M.S., Cadinu, D., Taurino, M., Piro, G. & Dalessandro, G. (2006). Antioxidant composition in cherry and high-pigment tomato cultivars. *Journal of the Agriculture and Food Chemistry*, Vol.54, pp. 2606-2613.

[42] Sass-Kiss, A., Kiss, J., Milotay, P., Kerek, M. M.& Toth-Markus, M. (2005). Differences in anthocyanin and carotenoid content of fruits and vegetables. *Food Research International*, Vol.38, pp.1023-1029.

[43] Brandt, S., Pék, Z., Barna, É., Lugasi, A. & Helyes, L. (2006). Lycopene content and colour of ripening tomatoes as affected by environmental conditions. *Journal of the Science Food and Agriculture, Vol.*86, pp.568-572.

[44] Sharma, S.K. & Le Maguer, M. (1996). Lycopene in tomatoes and tomato pulp fractions. *Italian Journal of Food Science. Vol.*2, pp.107-113.

[45] Toor, R. K. & Savage, G. P. (2005). Antioxidant activity in different fractions of tomatoes. *Food Research International.* Vol.38, pp.487-494.

[46] Shi, J. & Le Maguer, M. (2000). Lycopene in tomatoes: chemical and physical properties affected by food processing. *Critical Reviews in Food Science and Nutrition*, Vol.40,
pp.1-42.

[47] Al-Wandawi, H., Abdul-Rahman, M. & Al-Shaikhly, K. (1985). Tomato processing
waste an essential raw materials source. *Journal of Agriculture and Food Chemistry*, Vol.
33, pp.804-807.

[48] Fanasca, S., Colla, G., Maiani, G., Venneria, E., Rouphael, Y., Azzini, E. & Saccardo, F.
(2006). Changes in antioxidant content of tomato fruits in response to cultivar and
nutrient solution composition. *Journal of Agriculture and Food Chemistry*, Vol.54, pp.
4319-4325.

[49] Thompson, K.A., Marshall, M.R., Sims, C.A., Wei, C.I., Sargent, S.A. & Scott, J.W.
(2000). Cultivar, maturity and heat treatment on lycopene content in tomatoes. *Food
and Chemical Toxicology*, Vol.65, pp.791-795.

[50] Helyes, L., Brandt, S., Réti, K., Barna, É. & Lugasi, A. (2003). Appreciation and analysis of lycopene content of tomato. *Acta Horticulturae*, No.604, pp.531-537.

[51] Helyes, L., Dimény, J., Pék, Z. & Lugasi, A. (2006). Effect of maturity stage on content, color and quality of tomato (*Lycopersicon lycopersicum* (L.) Karsten) fruit. *Inerna-
tional Journal of Horticultural Science, Vol.*12, pp.41-44.

[52] Perkins-Veazie, P., Roberts, W. & Collins, J. K. (2006). Lycopene content among organically produced tomatoes. *Journal of Vegetable Science*, Vol.12, No.4, pp.93-106.

[53] Rossi, F., Godani, F., Bertuzzi, T., Trevisan, M., Ferrari, F. & Gatti, S. (2008). Health-
promoting substances and heavy metal content in tomatoes grown with different
farming techniques. *European Journal of Nutrition, Vol.47, pp.* 266-272.

[54] Borguini, R.G., Markowicz Bastos, D. H., Moita-Neto, J. M., Capasso, F. S. & da Silva
Torres, E.A.F. (2013). Antioxidant Potential of Tomatoes Cultivated in Organic and
Conventional Systems. *Brazilian Archives of Biology and Technology*, Vol.56, No.4, pp.
521-529.

[55] Vallverdú-Queralt, A., Jáuregui, O., Medina-Remón, A. & Lamuela-Raventós, R. M.
(2012). Evaluation of a method to characterize the phenolic profile of organic and
conventional tomatoes. *Journal of Agricultural and Food Chemistry*, Vol.60, No.13, pp.
3373-3380

[56] Oliveira, A.B., Moura, C.F.H., Gomes-Filho, E., Marco, C.A., Urban, L. & Miranda,
M.R A. (2013). The Impact of Organic Farming on Quality of Tomatoes Is Associated
to Increased Oxidative Stress during Fruit Development.*PLoS ONE, Vol* 8, No.2,
pp.e56354.

[57] Mitchell, A. E., Yun-Jeong, H., Koh, E., Barrett, D. M., Bryant, D. E., Denison, R. F. &
Kaffka, S. (2007). Ten-Year Comparison of the influence of organic and conventional

crop management practices on the content of flavonoids in tomatoes. *Journal of Agri-culre and Food Chemistry, Vol.* 55, No.15, pp.6154-6159.

[58] Kelly, S. D. & Bateman, A. S. (2010). Comparison of mineral concentrations in com-mercially grown organic and conventional crops – tomatoes (*Lycopersicon esculentum*) and lettuces (*Lactuca sativa*). *Food Chemistry, Vol.*119, pp.738-745.

[59] Ordonez-Santos, L. E., Vazquez-Oderiz, M. L. & Romero-Rodrýguez, M. A. (2011). Micronutrient contents in organic and conventional tomatoes (*Solanum lycopersicum L.*). *International Journal of Food Science and Technology, Vol.*46, No.1, pp.561-568.

[60] Ilić, S. Z., Kapoulas, N., Milenković, L. (2013). Micronutrient composition and quality characteristics of tomato from conventional and organic production. *Indian Journal of Agriculture Science,* Vol.83, No.6, pp.651-655.

[61] Gundersen, V., Mccall, D. & Bechmann, I.E. (2001). Comparison of major and trace element concentrations in Danish greenhouse tomatoes (*Lycopersicon esculentum* cv. Aromata F1) cultivated in different substrates. *Journal of Agriculture and Food Chemis-try,* Vol.49, pp.3808-3815.

[62] Rodrýguez, A., Ballesteros, A., Ciruelos, A., Barreiros, J. M. & Latorre, A. (2001). Sen-sory evaluation of fresh tomato from conventional, integrated, and organic produc-tion. *Acta Horticulturae, No.*542, pp.277-282.

[63] Dorais, M., Papadopoulos, A.P.& Gosselin, A.*(2001) Greenhouse tomato fruit quality. Journal of the American Society for Horticultural Science)* Vol.26, pp.239-319.

[64] Hernandez Suarez, M., Rodrýguez Rodrýguez, E. M. & Dýaz Romero, C. (2007). Min-eral and trace element concentrations in cultivars of tomatoes. *Food Chemistry, Vol.* 104, pp.489-99.

[65] Harker, R., Gunson, A. & Jaeger, S. (2003). The case of fruit quality: An interpretive review of consumer attitudes, and preferences for apples. *Postharvest Biology and Technology,* Vol.28, pp.333-347.

[66] Hauffman, S. & Bruce, Å. (2002). Matens kvalitet. Kungl Skogs-och Lantbruksakade-mien 1-15.

[67] Baldwin, E.A., Nisperos-Carriedo, M.O., Baker, R. & Scott, J. W. (1991). Qualitative analysis of flavor parameters in six Florida tomato cultivars. *Journal of Agriculture and Food Chemistry,* Vol.39, pp.1135-1140.

[68] Baldwin, E.A., Scott, J.W., Einstein, M.A., Malundo, T.M.M., Carr, B.T., Shewfelt, R.L. & Tandon, K.S. (1998). Relationship between sensory and instrumental analysis for tomato flavor. *Journal of American Society of Horticultural Science,* Vol.123, pp.906-915.

[69] Baldwin, E.A., Goodner, K., Plotto, A., Pritchett, K. & Einstein, M.A. (2004). Effect of volatiles and their concentration on perception of tomato descriptors. *Journal of Food Science,* Vol. 8, pp.310-318.

[70] Tandon, K.S., Baldwin, E.A. & Shewfelt, R.L. (2000). Aroma perception of individual volatile compounds in fresh tomatoes (*Lycopersicon esculentum*, Mill.) as affected by the medium of evaluation. *Postharvest Biololoy and Technology*, Vol.20, pp.261-268.

[71] Kapoulas, N., Ilić, S. Z., Milenković, L. & Mirecki, N. (2013). Effects of organic and conventional cultivation methods on micronutrient contents and taste parameter in tomato fruit. *Agriculture & Forestry*, Vol.59, No.3, pp.7-18.

[72] Gravel, V., Blok, W., Hallmann, E. *et al.* (2010). Differences in N uptake and fruit quality between organically and conventionally grown greenhouse tomatoes. *Agronomy for Sustainable Development, Vol.* 30, pp.797-806.

[73] Goyer, R. A. (1997). Toxic and essential metal interactions. *Annual Review of Nutrition, Vol.17*, pp.37-50.

[74] Hocking, P.J. & McLaughlin, M.J. (2000). Genotypic variation in cadmium accumulation by seed of linseed and comparison with seeds of some other crop species. *Australian Journal of Agriculture Research, Vol,* 51, pp.427-433.

[75] Puschenreiter, M., Horak, O., Friesl, W. & Hartl, W. (2005) Low-cost agricultural measures to reduce heavy metal transfer into the food chain – a review. *Plant Soil and Environment*, Vol.51, No.1, pp.1-11

[76] Jones, K.C. & Johnston, A.E. (1989). Cadmium in cereal grain and herbage from long-term experimental plots at Rothamsted, UK. *Environmental Pollution*, Vol.57, pp. 199-216.

[77] Santamaria, P., Elia, A., Serio, F. & Todaro, E. (1999). A survey of nitrate and oxalate content in retail fresh vegetables. *Journal of the Science Food and Agriculture, Vol.79*, pp. 1882-1888.

[78] Santamaria, P. (2006). Nitrate in vegetables: toxicity, content, intake and EC regulation. *Journal of the Science Food and Agriculture, Vol.86*, pp.10-17.

[79] Corre, W.J., Breimer, T. (1979). Nitrate and nitrite in vegetables, Pudoc, Wageningen, pp. 85.

[80] Maynard, D.N., Baker, A.V., Minotti, P.L., Peck, N.H. (1976). Nitrate accumulation in vegetables. *Advances in Agronomy, Vol.* 28, pp.71-118.

[81] Zhou, Z.-Y., Wang, M.-J. & Wang, J.-S. (2000). Nitrate and nitrite contamination in vegetables in China, *Food Reviews International*, Vol.16, pp.61-76.

[82] Grzebelus, D. & Baranski, R. (2001). Identification of accessions showing low nitrate accumulation in a germplasm collection of garden beet. *Acta Horticulrurae*, No.563, pp.253-255.

[83] Custic, M., Poljak, M., Coga, L., Cosic, T., Toth, N. & Pecina, M. (2003). The influence of organic and mineral fertilization on nutrient status, nitrate accumulation, and yield of head chicory. *Plant Soil and Environment, Vol.49*, pp.218-222.

[84] Basker, D. (1992). Comparison of taste quality between organically and conventionally grown fruits and vegetables. *American Journal of Alternative Agriculture.* Vol7, pp. 129-136.

[85] Rauter, W. & Wolkerstorfer, W. (1982). Nitrat in gemuse, Z. Lebensm. Unters. F. 175, 122-124.

[86] Pommer, G. & Lepschy, J. (1985). Investigation of the contents of winter wheat and carrots from different sources of production and marketing, Bayer. *Landwirtsch. Jahrb.* 62, 549-563.

[87] Hogstad, S., Risvik, E. & Steinsholt, K. (1997). Sensory quality and chemical composition of carrots: a multivariate study, *Acta Agriculturae Scandinavica,* Vol.47, pp. 253-264.

[88] Lairon, D., Spitz, N., Termine, E., Ribaud, P., Lafont, H. & Hauton, J.C. (1984). Effect of organic and mineral nitrogen fertilization on yield and nutritive value of butterhead lettuce, *Plant Foods for Human Nutrition,* Vol.34, pp.97-108.

[89] Lairon, D. (2010). Nutritional quality and safety of organic food. A review. *Agronomy for Sustainable Development,* Vol.30, No.1. pp.33-41.

[90] Umar, A.S. & Iqbal, M. (2007). Nitrate accumulation in plants, factors affecting the process, and human health implications. A review. Agronomy for Sustainable Development, Vol.27, pp.45-57.

7

Analysis of Production and Consumption of Organic Products in South Africa

Maggie Kisaka-Lwayo and Ajuruchukwu Obi

1. Introduction

Food and nutritional security remain an issue of global concern especially in developing countries. The practice of organic agriculture has been identified as a pathway to sustainable development and enhancing food security. Arguably, the most sustainable choice for agricultural development and food security is to increase total farm productivity *in situ*, in developing countries particularly sub-Saharan Africa. Attention must focus on the following: (i) the extent to which farmers can improve food production and raise incomes with low-cost, locally-available technologies and inputs (this is particularly important at times of very high fuel and agro-chemical prices); (ii) whether they can do this without causing further environmental damage; and (iii) the extent of farmers' ability to access markets [1]. Organic farming is one of the sustainable approaches to farming that can contribute to food and nutritional security [2]. Driven by increasing demand globally, organic agriculture has grown rapidly in the past decade [3]. Policy makers at the primary end of the food chains must wrestle with the dual objective of reducing poverty and increasing the flow of ecosystem services from rural areas occupied by small scale farmers and/or family farms [4].

Expectedly, a paradigm shift towards this realization of organic agriculture's role in food and nutritional security is emerging [5]. The United Nations Environmental Programme-United Nations Conference on Trade and Development, UNEP-UNCTAD [6] indicates that organic agriculture offers developing countries a wide range of economic, environmental, social and cultural benefits. On the development side, organic production is particularly well-suited for smallholder farmers, who make up the majority of the worlds' poor. Resource poor farmers are less dependent on external resources, experience higher yields on their farms and enjoy enhanced food security [7]. Organic agriculture in developing countries builds on and keeps alive their rich heritage of traditional knowledge and traditional land races. It has been observed to strengthen communities and give youth incentive to keep farming, thus reducing rural-urban migration. Farmers and their families and employees are no longer exposed to hazardous agro-chemicals, which is one of the leading causes of occupational injury and death in the world [7].

As organic production increases, so does the interest in organic market dynamics and studies are being carried out in order to analyse the future potential for organic agriculture. Figure 1 shows the global markets for certified organic products. In 2009, the global market for certified organic food and drink was estimated to be 54. 9 billion US dollars [8]. This represents a 37% growth from 2006 sales estimated at 40. 2billionUS dollars and a 207% increase from year 2000 sales estimated at17. 9 billion US dollars. In Africa, most of the organic farms are small family smallholdings [9] and certified organic production is mostly geared to products destined for export beyond Africa's shores. However, local markets for certified organic products are growing, especially in Egypt, South Africa, Uganda and Kenya [10]. Figure2 shows the ten countries in Africa with the largest proportion of land allocated to organic agriculture. South Africa has the third largest area under organic farming with 50, 000 hectares (ha), trailing Tunisia which has the largest area of 154, 793ha and Uganda with 88, 439ha [11]. Approximately 20% of the total area under certified organic farming in Africa is in South Africa, with 250 certified commercial farms [12]. With a few exceptions, notably Uganda, most African countries do not have data collection systems for organic farming and certified organic farming is relatively underdeveloped, even in comparison to other low-income continents. Some expert opinions suggest that this is due to lack of awareness, low-income levels, lack of local organic standards and other infrastructure for local market certification [13].

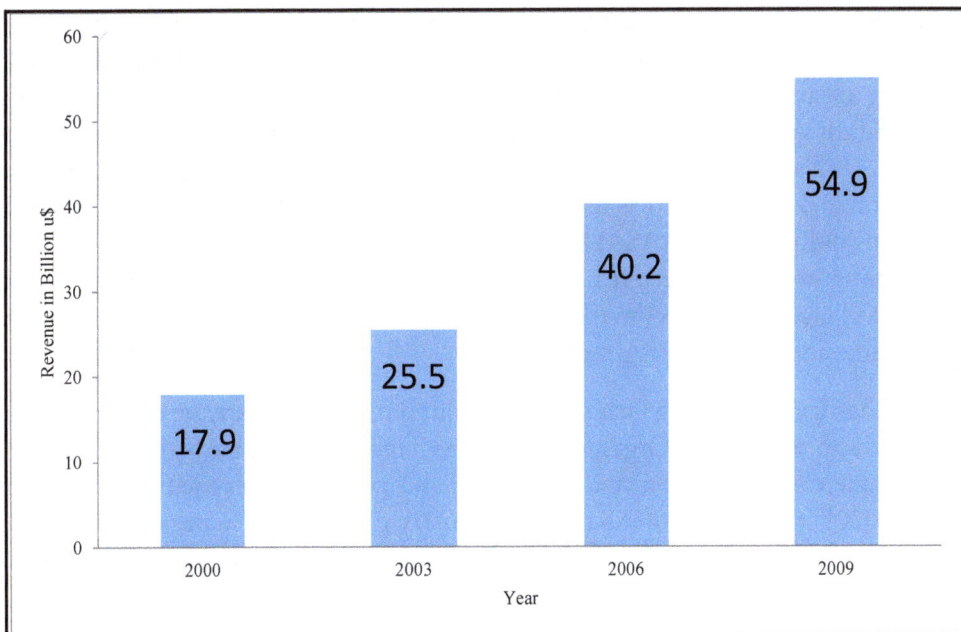

Figure 1. Development of the global market for organic products

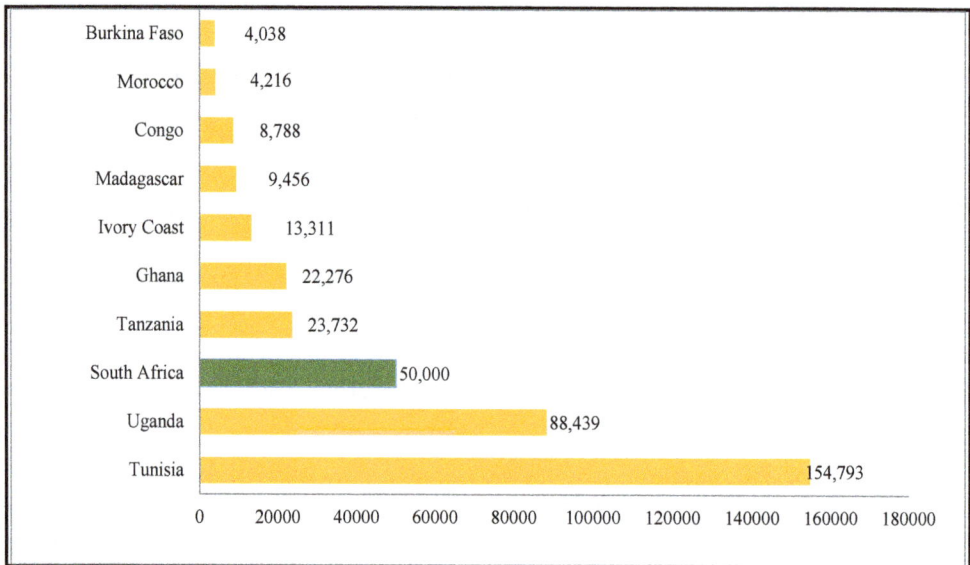

Source: [14]

Figure 2. The ten countries in Africa with the most organic agricultural land in hectares

2. Organic agriculture in South Africa

In 1999, only 35 farms were certified in South Africa, whereas in 2000 this number had increased to approximately 150 [15]. GROLINK [16] estimates that 240 farms with a total area of 43 620 ha (including pastures and in-conversion land) were certified in 2002. Certified organic produce in South Africa started with mangoes, avocadoes, herbs, spices, rooibos tea and vegetables [17]. This has now expanded to include a much wider range of products. Organic wines, olive oil and dairy products are now being produced [18]. The Organic Agricultural Association of South Africa (OAASA) estimates that there are approximately 100 non-certified farmers, farming about 1000hectares, following organic principles, who market informally through local villages or farmers markets (*ibid*). In the latter case, no differentiation is made between organic and non-organic produce.

South Africa has had an organic farming movement dating back many years, although it has grown in "fits and starts" [19]. Organic approaches have to make a trade off between market oriented commercial production and increasing the productive capacity of marginalized communities [20]. The growth of the organic industry has resulted in organic farming being practised in the Western Cape, KwaZulu-Natal, Eastern Cape, Northern Cape and Gauteng Province (Table 1). As discussed by [21] and [22] changing consumer preferences towards more health and environmental awareness has led to an increase in the demand for products produced using sustainable production methods. GROLINK [16] states that South Africa has in contrast with other Sub-Saharan countries, a substantial domestic market for organic products. This is an indication that the potential for organic farming in South Africa is not only based on access to the export market in Europe and the USA but also on the local demand. The domestic market is robust with two domestic retailers (Woolworth and Pick 'n' Pay) selling reasonable amounts of organic produce and both are now starting to insist on certification for this produce as well as farmers markets attracting large number of buyers.

One approach taken to improve smallholder access to organic markets has been the formation of certified organic groups using guidelines developed by the International Federation of Organic Agriculture Movement (IFOAM) and enforced by certification agencies such as Ecocert/AFRISCO (African Farmers Certified Organic) in the case of South Africa [23]. Under the group certification system, organic farmers can either grow and market their produce collectively or produce individually but market collectively. This ensures that smallholder farmers especially in developing countries are not marginalised and unduly excluded from the organic sector due to factors beyond their control. Several organic farming groups have emerged in South Africa in the last decade notably Ezemvelo Farmers Organization (EFO), Vukuzakhe Organic Farmers Organization (VOFO), Ikusasalethu Trust and Makhuluseni Organic Farmers Organisation.

The question of how to face the growing problem of food insecurity in Africa becomes more and more important, especially due to the steadily increasing world population and the changing consumption pattern. According to [24], while organically produced food seems not to be able to feed the World's Population, there are strong evidences that organic agriculture might help to alleviate the number of people suffering from hunger especially in developing countries. Given the strong negative externalities of conventional agriculture, the diversification of production as a basic principal of organic agriculture can contribute to the improvement of food security [25] which may improve the nutritional level in rural communities. The expanding global market for organic products [26, 27] and the possibilities for smallholder farmers in developing countries to access markets [24] can have very positive effects on the rural economies, triggering rural development. The increasing awareness of what people consume also has positive effects on organic agriculture as an alternative option for agricultural production. Organic agriculture may thus be an option in some areas to strongly support rural development.

Against this background, the objective of this paper is to provide, through an exploratory analysis of data from farm and households surveys, empirical insights into determinants of organic farming adoption, differentiating between fully-certified organic, partially-certified organic and non-organic farmers; eliciting farmers risk preferences and management strategies and; exploring consumer awareness, perceptions and consumption decisions. By exploring a combination of adoption relevant factors in the context of real and important land management choices, the paper provides an empirical contribution to the adoption literature and provides valuable pointers for the design of effective and efficient public policy for on-farm conservation activities. Similarly, achieving awareness and understanding the linkage between awareness and purchasing organics is fundamental to impacting the demand for organically grown products. Consumer awareness of organic foods is the first step in developing demand for organic. Section 3 describes the materials and methods, outlining the study areas and study methodology. Section 4 presents the results and discussion. Finally, section5 provides concluding remarks.

3. Materials and methods

The study was carried out in the two provinces of KwaZulu-Natal and the Eastern Cape Provinces in South Africa (Figure 3). The selected study areas are in the rural Umbumbulu Magisterial District in KwaZulu-Natal Province and the OR Tambo and Amatole District Municipalities in the Eastern Cape Province.

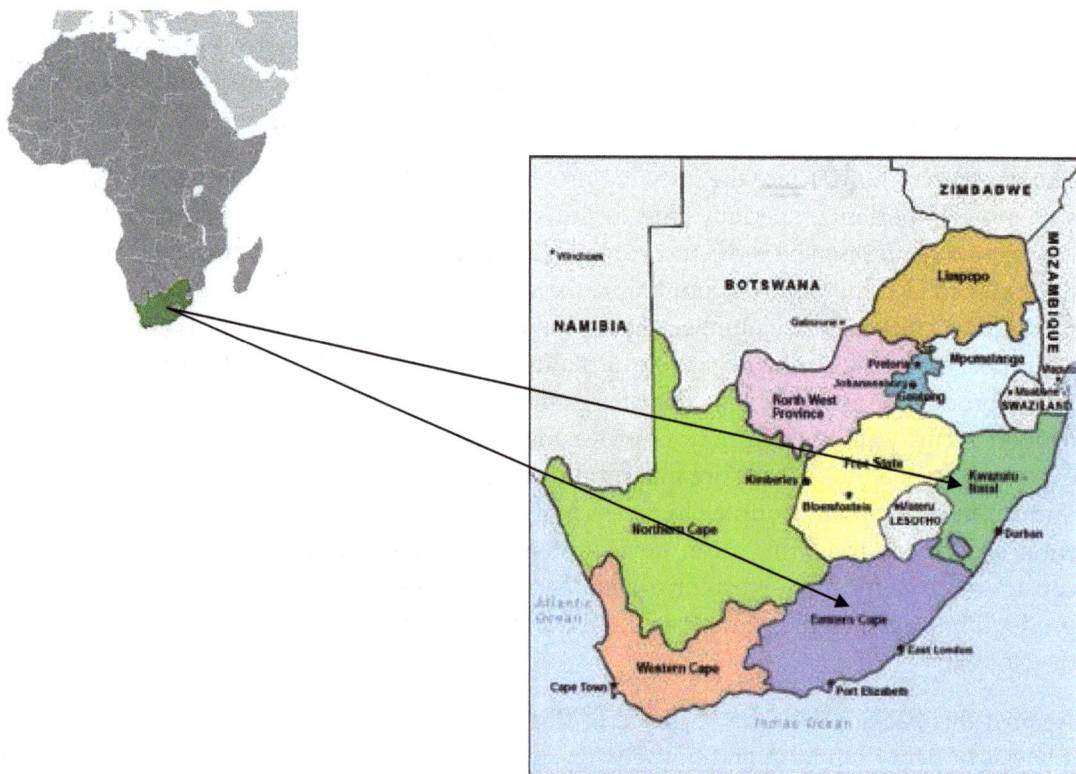

Figure 3. Map of study area

The Umbumbulu area is one of the former homelands of KwaZulu-Natal Province. The Province has the largest concentration of people who are relatively poor, and social indicators point to below average levels of social development. According to the mid-year population estimates by Statistics South Africa [28], the Province has a population of 10. 6 Million people 67 percent of whom reside in communal areas of the former KwaZulu-Natal homeland [28]. The OR Tambo District Municipality is the second poorest Municipality in the Eastern Cape Province with some areas having poverty levels of as high as 82 % [29]. About 67% of the households within the district have income levels that range between R0 and R6, 000. The District Municipality has the second highest population of all the districts with more than 1, 504, 411 inhabitants [29]. For a mostly rural district it also has a high population density of 90 people per square kilometre. The Amatole District Municipality is named after the legendary Amatole Mountains and is the most diverse District Municipality in the Province. Two-thirds of the District is made up of ex-homeland areas. The District has a moderate Human Development Index of 0. 52 with over 1, 635, 433 inhabitants [30], and a moderately high population density of 78 people per square kilometre. The population is mainly African with some whites and coloureds. Amatole District Municipality has the second highest economy in the province.

The Eastern Cape Province is bordering KwaZulu-Natal with similarities in the socio-economic status and rurality of the two Provinces. Both Provinces' economic dependence is on agriculture with huge potential for organic agriculture development. The Eastern Cape is also a major consumer of produce from KwaZulu-Natal. A total of 400 respondents were interviewed, representing 200 farmer respondents from KwaZulu-Natal and 200 consumer respondents from Eastern Cape Provinces. The survey farmers in Umbumbulu District, KwaZulu-Natal were stratified into three groups: fully-certified organic farmers, partially-certified organic farmers and non-organic farmers. While the 48 fully-certified farmers and 103 partially-certified farmers were purposively selected, the sample of 49 non-organic farmers was randomly selected within the same region from a sample frame constructed from each of the five neighbouring wards. The survey was conducted by a team of trained enumerators from the study area. These enumerators had to be fluent in both English and Zulu. A questionnaire was used to record all household activities (farm and non-farm), enterprise types, crop areas and production levels, inputs, expenditures and sales for the past season. The questionnaires also captured socio-economic and institution data such as household characteristics, land size and tenure arrangements, farm characteristics and investment in assets. Other questions related to farmers' management capacity and demographic characteristics such as the supply of on-farm family labour and education status.

The farmers' risk attitude was elicited using the experimental gambling approach as outlined by [31]. Here, the study farmers were presented with a series of choices among sets of alternative prospects (gambles) that do not involve real money payments. Respondents were required to make a simple choice among eight gambles whose outcomes were determined by a flip of a coin. The experimental approach remedies some of the more serious measurement flaws of the direct elicitation utility (DEU) interview method reporting that evidence on risk aversion using direct elicitation utility through pure interviews is unreliable, nonreplicable and misleading even if one is interested only in a distribution of risk aversion rather than reliable individual measurements [31, 32]. The farmers were further asked in the field survey to give their perceptions of the main sources of risk that affect their farming activity by ranking a set of 20 potential sources of risk on like rt-type scales ranging from 1 (no problem) to 3 (severe problem). These sources of risk were developed from findings of the research survey and from past research on the sources of risk in agriculture, challenges that smallholder farmers face in trying to access formal supply chains. The farmers were also requested to score any other sources of risk(s) that they wanted to add to the list of hypothesized sources of risk. These sources of risk are ranked from 1 (being the most important source of risk) to 20(being the least important source of risk ones). The ranking was done by averaging the scores on each source of risk and assigning a rank accordingly.

The study area in the Eastern Cape was stratified into the OR Tambo District Municipality and the Amatole District Municipality representing a broad spectrum of consumers across the Province. The stratified study areas were further clustered into rural, peri-urban and urban areas. The respondents were selected by simple random sampling to avoid bias. A total of 100 consumers were selected from OR Tambo District Municipality and represented by a selection of 30 respondents from peri-urban location, 40 respondents from urban suburbs and 30 respondents from rural areas. In the Amatole District Municipality, 100 consumers selected and interviewed included 30 respondents from rural Cata, 40 urban respondents from the East London Suburbs and lastly 30 respondents drawn from the peri urban area of Kwezana and Tsathu villages. A structured questionnaire was used that covered the respondent's socio-economic and demographic background, consumer knowledge and awareness of organic products, perceptions, attitudes as well as consumption decisions.

The ordered probit model was used to identify the determinants of farmers' decision to participate in organic farming. The dependent variable is the farmer's organic farming status and was placed in three ordered categories in the survey. The model is estimated as:

$$\text{Organic farming status} = f \begin{pmatrix} age, \ gender, \ education, \ household \ size, \ farm \ size, \ farm \ income, \ off \ farm \ income, \ input \ costs, \\ land \ tenure, \ location, \ land \ tenure, \ livestock, \ chicken \ ownership, \ risk \ attitudes \ and \ assets \end{pmatrix} \quad (1)$$

The organic farming status is modelled using the ordered probit model with the model outcomes:

S_i=3 (fully-certified organic),

S_i=2 (partially-certified organic farming) and

S_i=1 (non-organic farmers).

The farmer's decision on their organic farming status is unobserved and is denoted by the latent variable s_i^*. The latent equation below models how s_i^* varies with personal characteristics and is represented as:

$$s_i^* = X_i'\alpha + \varepsilon_i \quad (2)$$

Where:

- the latent variable s_i^* measures the difference in utility derived by individual i from either being fully-certified organic, partially-certified organic or non-organic.

- (i=1, 2, 3................ n) n represents the total number of respondents. Each individual i belongs to one of the three groups.

- X_i is a vector of exogenous variables,

- α is a conformable parameter vector, and

- the error term ε_i is independent and identically distributed as standard normal, that is $\varepsilon_i \sim NID\ (0, 1)$.

The observed variable (S_i) relates to the latent variable (s_i^*) such that

$$
\begin{aligned}
S_i = \ &1 \quad if\ s_i^* \leq 0 \\
&2 \quad if\ 0 < s_i^* > \gamma \\
&3 \quad if\ s_i^* > \gamma
\end{aligned}
\tag{3}
$$

Taking the value of 3 if the individual was fully-certified organic and 1 if the individual was non-organic. The implied probabilities are obtained as:

$$
\begin{aligned}
Pr\ \{S_i = 1 \mid X_i\} &= Pr\ \{s_i^* \leq 0 \mid X_i\} = \Phi(-X_i'\alpha) \\
Pr\ \{S_i = 3 \mid X_i\} &= Pr\ \{s_i^* > \gamma \mid X_i\} = 1 - \Phi(\gamma - X_i'\alpha) \\
&\text{and} \\
Pr\ \{S_i = 2 \mid X_i\} &= \Phi(\gamma - X_i'\alpha) - \Phi(-X_i'\alpha)
\end{aligned}
\tag{4}
$$

Where γ is the unknown parameter that is estimated jointly with α. Estimation is based upon the maximum likelihood where the above probabilities enter the likelihood function. The interpretation of the α coefficients is in terms of the underlying latent variable model in equation 11.

The probability of the farmer being fully-certified organic can be written as

$$
Pr(S_i = 1) = \Phi(X_i'\alpha_1),
\tag{5}
$$

Where $\Phi(\ )$ is the cumulative distribution function (cdf) of the standard normal [33].

A measure of goodness of fit can be obtained by calculating

$$
\rho^2 = 1 - \left[lnL_b / lnL_o \right]
\tag{6}
$$

Where lnL_b is the log likelihood at convergence and lnL_o is the log likelihood computed at zero. This measure is bounded by zero and one. If all model coefficients are zero, then the measure is zero. Although ϱ^2 cannot equal one, a value close to one indicates a very good fit. As the model fit improves, ϱ^2 increases. However the ϱ^2 values between zero and one do not have a natural interpretation [34]. Another similar informal goodness of fit measure that corrects for the number of parameters estimated is

$$\rho^2 bar = 1 - \left[lnL_b K \,/\, lnL_o \right]. \tag{7}$$

Where K is the number of parameter estimates in the model (degrees of freedom)

For the experimental gambling approach, the utility function with Constant Partial Risk Aversion (CPRA) is used to get a unique measure of partial risk aversion coefficient for each game level. This depicted as the equation below:

$$U = (1 - S)c^{(1-S)}. \tag{8}$$

Where

S=coefficient of risk aversion, and

c=certainty equivalent of a prospect.

The Herfindahl Index (DHI) is used to calculate enterprise diversification and represent the specialization variable. Although, this index is mainly used in the marketing industry to analyze market concentration, it has also been used to represent crop diversification [35, 36]. Herfindhal index (DHI) is the sum of square of the proportion of individual activities in a portfolio. With an increase in diversification, the sum of square of the proportion of activities decreases, so also the indices. In this way, it is an inverse measure of diversification, since the Herfindhal index decreases with an increase in diversification. The Herfindhal index is bound by zero (complete diversification) to one (complete specialization).

$$\text{Herfindhal index} \left(\text{DHI} \right) = \sum_{i=1}^{N} s_i^2 \tag{9}$$

Where

N=number of enterprises and

s_i=value share of each *i-th* farm enterprise in the farm's output. $s_i = {x_i}\Big/{\sum_1 x_i}$ is the proportion of the *i-th* activity in acreage / income.

4. Results and discussions

4.1. Determinants of adoption of organic farming

The summary statistics in Table 1 show that the average age of the farmers was over 50 years with younger people migrating to urban centres in search of better jobs. In the study area, most of the men are engaged in wage employment at the neighbouring sugarcane farms or as migrant workers in the cities of Durban, Johannesburg. Hence over 70% of the farmers were female. Education levels are low and are consistent with most rural farming communities in South Africa, where formal education opportunities are limited. Household sizes were large with family labour playing a major role in tilling the land. Small farm sizes averaging 0. 59 hectares for fully-certified organic farmers, 0. 67 hectares for non-organic farmers and 0. 71 hectares for partially-certified farming was common in KwaZulu-Natal.

The main sources of income were farm and off farm employment, the latter constituting wages or salary income and remittances. Farm income was highest for fully-certified organic farmers. This is an indication that the adoption of fully-certified organic farming and its commercialization has brought economic benefits to these otherwise poor rural households and is an important contributor to household income. The proportion of income from farming was highest among the fully certified organic farmers. While the average farmer was classified as risk averse, non-organic farmers were more risk averse than their organic counterparts. Risk-averse farmers are reluctant to invest in innovations of which they have little first-hand experience. Despite the tenure system being communal, farmers felt they had tenure rights through the permission to occupywith allocation done by the traditional chief of the tribe (*inkosi*) and his headman (*induna*). On average the farmers acknowledged that the household had rights to exercise on its own cropland the building of structures, planting trees and bequeathing to family members or leasing out. Fully certified farmers had more assets than their non-organic counterparts as well as chicken and livestock.

Variable	Fully-certified organic		Partially-certified organic		Non-organic	
	Mean	Std. Dev	Mean	Std. Dev	Mean	Std. Dev
Age (years)	52.60	1.90	48.60	1.41	52.70	2.11
Gender (1=female)	0.82	0.05	0.71	0.05	0.84	0.05
Education (years)	4.94	4.24	4.37	4.49	3.38	0.61
Household size	9.49	5.23	7.72	3.68	6.60	3.46
Land size (hectares)	0.59	1.22	0.71	1.16	0.67	1.43
Input costs (rand/year)	812.90	884.90	309.30	343.40	318.20	302.90
Proportion of income from farming	0.62	0.79	0.38	1.04	0.39	0.63
Farm income (rands/year)	973.17	1074.51	417.26	271.50	400.53	429.53
Location	2.56	0.60	1.91	0.54	4.00	0.00
Arrow Pratt Risk Aversion coefficient	0.55	0.29	0.58	0.31	0.76	0.29
Land rights (0 = no)	1.98	0.14	1.75	0.56	1.93	0.33
Chicken ownership	15.29	13.16	9.25	8.69	6.40	6.62
Asset ownership (index)	0.98	0.60	0.56	0.59	0.67	0.75

Table 1. Summary statistics of sampled farmers in KwaZulu-Natal (n=200)

The ordered probit model results are presented in Table 2. The model successfully estimated the significant variables associated with the farmer's adoption decisions. The Huber/White/ sandwich variances estimator was used to correct for heteroscedasticity. The explanatory variables collectively influence the farmer's decision to be a certified organic with the chi-square value significant at one percent. The following variables were found to be significant determinants in the organic farming adoption decision by smallholder farmers in the study area: age, household size, land size, locational setting of the farmer depicted by the sub-wards Ogagwini, Ezigani, and Hwayi, farmer's risk attitude, livestock ownership (chicken and goat ownership), land tenure security as depicted by the rights the farmer can exercise on his/her own cropland to build structures and asset ownership.

Variables	Parameter	Robust std error	P-values
Age	0.0194072	0.0079204	0.014***
Gender	0.3796234	0.2707705	0.161
Household size	0.0504668	0.0271520	0.063*
Land size	-0.2352607	0.1083583	0.030**
Off Farm Income	-0.0001223	0.0001129	0.279
Location (sub-ward)			
Location (1= ogagwini)	2.894311	0.6380815	0.000***
Location (1=ezigani)	4.191274	0.7234394	0.000***
Location (1=hwayi)	5.158803	0.8495047	0.000***
Risk attitudes	-0.759508	0.3773067	0.044**
Fertility (Manure)			
Chicken ownership	0.0424046	0.0148472	0.004***
Cattle ownership	-0.0418692	0.0431078	0.331
Goat ownership	-0.1005212	0.0569375	0.077*
Land tenure rights			
Land tenure (1= build structures)	0.4803418	0.2372247	0.043**
Land tenure (1= plant trees)	0.0235946	0.3023182	0.938
Land tenure (1= bequeath)	0.1335225	0.2619669	0.610
Land tenure (1= lease out)	-0.3840883	0.2593139	0.139
Land tenure (1= sell land)	0.0829177	0.2978485	0.781
Asset ownership	0.5853967	0.205389	0.004***

*Significance levels: *** p<0. 01: ** p<0. 05: * p<0. 1*

(Source: Field Data)

Table 2. Adoption of organic farming among smallholder farmers: Ordered probit model results

The study established that older female farmers with large household sizes were more likely to be certified-organic. Similarly, farmers who reside in the sub-wards Ogagwini, Ezigani, and Hwayi were more likely to be certified organic. This suggests the presence of local synergies in adoption which raises the question about the extent to which ignoring these influences biases policy conclusions. The negative correlation between land size and adoption implies that smaller farms appear to have greater propensity for adoption of certified organic farming. This finding is supported by several studies reviewed in the literature that allude to the fact that organic farms tend to be smaller than conventional farms. The significance of livestock is explained by the importance of manure for organic farming. The study also found that older farmers tend to be adopters supporting findings by [37]. The propensity to adopt was also positively influenced by asset index which is a proxy for wealth.

4.2. Risk aversion and risk management strategies

The distribution of risk aversion preferences for each prospect for the fully-certified organic, partially-certified organic and non-organic crop farmers are presented in Table 3. The distribution of responses was spread across all classes of risk aversion for the pooled data. It can be noted that on average, the majority of the respondents revealed their preference for prospects representing intermediate and moderate risk aversion alternatives across the three farmer groups. Table 3 further shows that non-organic farmers were the most risk averse being classified as extremely risk averse at 20. 4%, compared to fully and partially-certified organic farmers at 7. 3% and 4. 2%, respectively. This explains their non-adoption of certified organic farming, despite its introduction in the area since the year 2000. On the other hand, the fully-certified organic farmers were the least risk averse, being classified as neutral to risk preferring at 9. 1% compared to 7. 3% and 4. 1% for the partially certified and non-organic farmers respectively. These results conform to *a priori* expectations regarding the risk preference patterns of smallholder farmers.

Farmer group	Risk aversion classification					
	Extreme	Severe	Intermediate	Moderate	Slight to neutral	Neutral to preferring
Fully certified organic (n = 48)	7.30	5.50	30.90	40.00	7.30	9.10
Partially certified organic(n = 95)	4.20	8.30	44.80	29.20	5.20	7.30
Non-organic (n= 46)	20.40	8.20	30.60	30.60	0.00	4.10
Pooled data (n = 189)	9.00	7.50	37.50	32.50	4.50	7.00

Source: Field data

Table 3. Distribution of smallholder farmers according to risk preference patterns in KwaZulu-Natal

According to Figure 4, the non-organic farmers constituted 55. 6% of respondents within the extreme risk aversion class compared to 22. 2% for fully-certified organic and 22. 2% for partially-certified organic farmers. This is a confirmation of previous findings in this study that explains the non-adoption of certified organic farming by the non-organic farmers. In the risk neutral to preferring category, the non-organic farmers constitute only 14. 3%. Fully-certified organic farmers constituted 57. 1% and partially-certified organic farmers constituted 28. 6%.

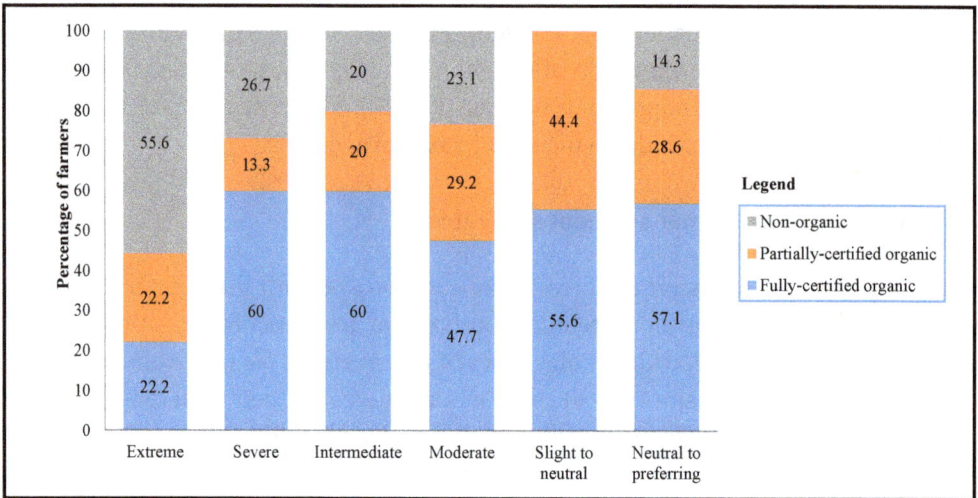

Figure 4. Frequency distribution within risk aversion classes across the farmer groups

A comparison of the results from the South African study, which applied the general experimental method, with similar studies using the same methodology was for farming communities in the Côte d'Ivoire [38], Ethiopia [39], Zambia [40], Philippines [41] and India [31], shows similarities in the findings of the studies done in India, Philippines, Zambia and Côte d'Ivoire, where the majority of the respondents are classified as intermediate to moderate risk aversion (Table 4). Similarly, these results suggest that farm households in South Africa are less risk averse than in Ethiopia, Zambia and Côte d'Ivoire but are much more risk averse than in India and Philippines.

Farmers identified their sources of risk and significance in terms of the potential impact to their farming activity as presented in Table 5. The fully-certified organic farmers cited in order of priority, uncertain climate (mean 2. 96), lack of cash and credit to finance inputs (mean 2. 78) and tractor unavailability when needed (mean 2. 76). These risk sources have a direct bearing on production of organic produce. Climatic conditions are beyond the farmers' control, and the top ranking probably reflects the farmers' concerns about the effects of recent drought in the Umbumbulu district. These impacts negatively on crop yield. Due to communal land ownership and strict conditions for credit, farmers have limited options to obtain production credit from financial institutions. Among the sampled farmers only 21 farmers were able to access credit. Farmers in the study area lack collateral that is acceptable to banks. For example, banks required title deeds as proof of land ownership but the majority of black farmers in

Studies	Extreme to severe risk aversion	Intermediate to moderate risk aversion	Risk-neutral to risk preferring	Number of responses
India [31]				
50 rupee	8.4	**82.2**	9.4	107
500 rupee	16.5	**82.6**	0.9	115
Philippines [41]				
50peso	10.2	**73.5**	16.3	49
500peso	8.1	**77.6**	14.3	49
Zambia [40]				
1000kw	29.1	**46.4**	24.5	423
10000kw	36.7	**52.5**	11	137
Ethiopia [39]				
5bir	**45.4**	33.6	21	262
15bir	**55.7**	27.5	16.8	262
Côte d'Ivoire [38]				
1000FCFA	32.8	**53.9**	13.3	362
5000FCFA	**46.1**	45.9	8	362
*South Africa [42]				
400Rands	16.5	**70**	11.5	196

*Source: Field work

Table 4. Percentage distribution of revealed risk preferences in five experimental studies

South Africa and especially in the former homelands still lacked this vital documentation. Tractor unavailability can be attributed to the fact that there is one tractor that has been allocated to the members of Ezemvelo Farmers Organisation. The tractor is leased out at a rental fees. This poses a challenge during the land preparation phase when the demand for its services is at peak.

Similarly, partially-certified farmers also ranked tractor not being available when needed (mean 2. 89) and uncertain climate (mean 2. 83) as identified sources of risk (Table 5). The risk of delays in payment for products sent to pack house (mean 2. 89) are attributed to various factors, among them the contractual obligation the agent has with the retailer which has a bearing on the duration of payment. Payment is only made to the farmer once the supply has been forwarded to the retailer and there is confirmation of the quantity of produce that has been rejected. The process flow delays payments to farmers. Non-organic farmers also cited uncertain climate (mean 2. 82), livestock damage to crops (mean 2. 80) and lack of cash and credit to finance farm inputs (mean 2. 78). The livestock damage is a result of lack of fencing around the crops planted.

Constraint	Fully-certified organic			Partially-certified organic			Non-organic		
	Mean	Std.Dev.	Rank	Mean	Std.Dev.	Rank	Mean	Std.Dev.	Rank
Livestock damage	2.56	0.774	7	2.82	0.448	4	2.8	0.539	2
Uncertain climate	2.96	0.189	1	2.83	0.409	3	2.82	0.486	1
Uncertain prices for products sold to pack house	2.21	0.793	13	2.13	0.591	16	-	-	-
Uncertain prices for products sold to other markets	1.94	0.811	17	2.02	0.595	18	2.17	0.761	10
Huge work load	2.58	0.599	6	2.32	0.688	12	2.53	0.649	4
Lack of cash and credit to finance inputs	2.78	0.567	2	2.58	0.615	6	2.78	0.468	3
Lack of information about producing organic crops	2.02	0.687	15	2.2	0.632	14	2.16	0.717	11
Lack of information about alternative markets	2.38	0.623	10	2.29	0.602	13	-	-	-
Lack of proper storage facilities	2.56	0.66	7	2.46	0.543	9	2.41	0.643	7
Lack of affordable transport for products	2.72	0.492	4	2.42	0.56	11	2.06	0.852	12
Lack of telephone to negotiate sales	2.69	0.509	5	2.55	0.633	8	2.22	0.771	8
Inputs not available at affordable prices	2.52	0.642	9	2.8	0.447	5	2.51	0.545	5
Tractor not available when needed	2.76	0.501	3	2.89	0.416	1	2.46	0.713	6
Cannot find manure for purchase	1.92	0.778	18	2.56	0.66	7	2.2	0.645	8
Cannot find labour to hire	1.73	0.764	20	1.76	0.816	20	2	0.764	13
Cannot access more cop land	1.95	0.753	16	1.98	0.805	19	1.92	0.794	14
Delay of payment of products sent to pack house	2.22	0.723	12	2.89	0.315	1	-	-	-
Lack of bargaining power over product prices at the pack house	2.16	0.672	14	2.2	0.704	14	-	-	-
Lack of information about consumer preferences for organic products	2.23	0.654	11	2.44	0.604	10	-	-	-
No reward system or incentive for smallholder producers	1.86	0.78	19	2.02	0.866	17	-	-	-

Table 5. Identification and ranking of risk sources by farmers

The most important traditional risk management strategies used by the farmers were identified as crop diversification, precautionary savings and participating in social network. The overall Herfindahl index of crop diversification is estimated at 0. 61 which indicates that the cropping system is relatively diverse (Table 6). These results confirm previous findings by [43] who obtained an estimated DHI of 0. 49-0. 69 among smallholder farmers in three regions in Bangladesh. As shown in Table 6, non-organic farmers practiced more crop diversification with a DH index of 0. 23 compared to organic farmers with a DHI of 0. 72. These results are consistent with previous findings in this study measuring farmers risk attitudes and presented in Figure 6. 8, that established that smallholder farmers in the study area tend to diversify due to their risk averse nature and that non-organic farmers are more risk averse than organic farmers.

According to Table 6, a total of 69. 1% of fully-certified farmers practised crop diversification compared to 96. 8% of the non-organic farmers. A total of 81. 2% of the partially certified farmers practised crop diversification. The common crops grown by the organic farmers are amadumbe, potatoes, sweet potatoes and green beans while non-organic farmers grew amadumbe, potatoes, sweet potatoes, green beans, maize, sugarcane, bananas, chillies and peas.

No.	Risk management strategy	Fully-certified organic	Partially-certified organic	Non-organic
		n = 48	n = 103	n = 49
1	Enterprise diversification index (DH)	0.72	0.89	0.23
2	Practice crop diversification (% of respondents)	69.10	81.20	96.80
3	Savings bank account (% of respondents)	60.90	48.90	46.80
4	Current level of savings (% of respondents)			
	less than R500	27.27	37.84	35.29
	R501 – R1000	45.45	29.73	41.18
	R1001 – R5000	21.21	29.73	17.65
	More than 5000	6.07	2.70	5.88
5	Social networks (% of respondents)			
	Membership of EFO	100.00	100.00	10.00
	Others (burial clubs, *stockvels*)	33.00	25.00	25.00

Table 6. Risk management strategies used by the different farmer groups

Precautionary saving occurs in response to risk and uncertainty [44]. The smallholder farmers' precautionary motive was to delay/minimise consumption and save in the current period due to their lack of crop insurance markets. According to [45], the quantitative significance of precautionary saving depends on how much risk consumers face. Whereas 60. 9% of the fully certified farmers had savings bank accounts, only 46. 8% non-organic farmers had bank accounts. The current level of saving in the study area was low with savings ranging from less than R500 to over R5000 per month. The level of savings was low across all groups. Among the fully-certified organic group, most of the respondents (45. 45%) saved between R1000-R5001 whereas most of the partially-certified farmers (37. 84%) saved less than R500 per month. Most of the non-organic farmers (41. 18%) saved between R501-R1000 per month. Across all groups, however the level of saving greater than R5000 was minimal.

The farmers also engage in social networks as a risk sharing strategy. There were two main categories of social networks that the farmers engaged in. These are farmers association and other social networks most notably burial clubs and stockvels. The farmers association is used as a vehicle by the organic farmers to gain access to markets for their organic produce while the burial clubs and stockvels are sources of access to credit and/or loans. In the latter instance, farmers do not have to produce collateral. The burial clubs and stockvels are common in most rural areas and are a source of mitigating liquidity and financial risk where possible.

4.3. Consumer awareness, perceptions and consumption decisions

The summary statistics of consumers presented in Table 7 showed that the majority of the consumers were females within 25-34 age category. Previous studies for example [46] found that women were the predominant purchasers of organic food and responsible for household consumption. The younger generation consumers represent an important target group in the advancement of consumer demand for organic products. The level of education was generally low especially among rural consumers. The unemployment rates in the former homelands demonstrates a substantial skewering of the demographic profile of the district and high dependency rates of those not economically and productively active. It also reflects the levels of out migration of economically active population from the province to other parts of South Africa. Unemployment was also lower in urban areas than rural areas. The income distribution of the respondents is especially concentrated in the R1000 – R5000/month category. However the majority of the respondents within this category were in the rural areas. This can be attributed to limited economic activity in rural areas. The household size was within the provincial estimate of 4-5 persons per household [47] with rural households having higher numbers. Majority of the respondents had children under the age of 18 years in the household. The average distance to the nearest shops were estimated at between 6-9kms. In the urban areas however this was reduced to 1. 38kms.

There is a general understanding of term 'organic foods' among consumers. Consumers defined organic foods as healthy and nutritious, associated with traditional and or indigenous methods of production and free from chemicals. There were low levels of awareness about local standards for organic products, the identification of organic products using an organic logo, existence of a national organic movement and/or the presence of an organic certification body in South Africa. Therefore consumers could not readily identify certified organic against non-certified organic products. Notwithstanding, consumers argued that there was a need for certification and verification of organic products and hence are unable to make informed decisions on the organic status of products in the market.

Trust of organic labels can be increased once more information is available to consumers on the various organic labels, their meaning and on the difference between certified and non

Variable		Former Homelands		Locality		
		OR Tambo DM	Amatole DM	Rural	Peri-urban	Urban
		n = 100	n = 100	n = 30	n = 30	n = 40
Gender	Male	43	34	28	40	44
	Female	57	66	72	60	56
Age in years	18-24	17	13	18	16	12
	25-34	29	33	12	34	26
	35-44	27	16	14	23	41
	45-55	20	17	21	19	16
	>55	7	21	35	8	5
Education Level	None	4	9.7	16.1	6.5	1.2
	Primary	21	29.1	46.4	32.3	5.9
	High school	39	39.8	37.5	48.4	34.1
	Tertiary	36	21.4	0	12.9	58.8
Employment Status	Unemployed	29.4	31	52.6	48.3	2.4
	Student	9.8	4	5.3	5	9.4
	Housewife/man	10.8	8	19.3	10	2.4
	Retired	5.9	1	8.8	1.7	1.2
	Working part-time	14.7	11	8.7	18.3	11.8
	Working full time	29.4	45	5.3	16.7	72.8
Income Level	<1000	10	12.5	0	4.8	23.5
	1001 – 5000	16	5.8	0	0	25.9
	5001-10 000	20	17.3	5.3	17.7	28.2
	10 001 – 15 000	30	49	66.7	46.8	16.5
	>15 000	24	15.4	28.1	30.6	5.9
Household size in number		5.2	4.33	5.18	4.98	4.31
Children < 18 years		79	55.8	71.9	71	61.2
Distance in Kms		6.71	9.63	12.67	9.32	1.38

Table 7. Summary statistics of consumers in the Eastern Cape Province

certified products in the shelves. In the absence of this information, producers and likewise consumers may not get value for money. Certification and labelling is essentially in regulating and facilitating the sale of organic products to consumers. The perception of the high price of organic products is a deterrent to the purchase of organic products and hence the growth of organic industry especially for the emerging organic market of South Africa. To increase the consumption of organic products, it will be important to motivate new consumer segments to buy organic food. Hence trust is a crucial aspect when consumers decide whether to buy or not to buy organic products [48].

Trust is a 'credence attribute' which is not directly observable by consumers. Enhancing consumers trust about the labels of organic products can be achieved through among others, effective communication strategies on the traceability of organic products and ensuring compliance and adherence by retailers selling organic products to the certification standards and availability of information on the organic status of products. Some of the reasons advanced in the study to increase consumers trust for organic products is to:

- purchase from specific shops that sell organic

- check for organic certification label

- practice own organic farming

In South Africa, food retailers have the largest share of the organic industry [49]. Similarly, most products are sold through the export market due to the higher revenue from exports. Irwin [50] says that South Africa has a favourable position for expansion in the domestic market as a result of the following developments in the organic sector over the past few years:

- establishment of separate organic section in major retail stores

- national regulation/standards for organic products

- establishment of South Africa organic certification bodies

- formation of South African organic associations.

Food purchasing is an important part of food behaviours. In this study the apportioning, explicitly or tacitly, of the responsibility of household food shopping depends on a number of factors as food purchasing is an important part of food behaviours. This responsibility was closely shared among various members of the household with majority of the consumers being responsible for the decision making of organic food demand and purchase. The general finding in the study was that most consumers shop in supermarkets, grocery stores and *spaza (kiosks)* shops. The majority of consumers who shop in supermarkets reported that local shops do not provide the services people demand and that food choice and quality are limited. This is coupled with discount promotions common with supermarkets and variety of products. The findings from this study are consistent with findings from the Food Safety Agency [51] that state that a vast majority (92%) of consumers continue to use supermarkets for most of their food shopping. However, local shops play an important role in 'top-up' shopping, being used by 75% consumers for some of their food purchases.

Commonly consumed organic products included fresh vegetables, fresh fruits, meat/meat products and milk/ milk products. However, the general trend in Figure 5 and Figure 6 shows that there are marked increases in the future demand of all organic products. This augurs well for the growth of the organic industry in the Eastern Cape and in South Africa in General. The findings of this study are consistent with [52] who stated that a study by Pick-n-Pay, one of the major national retail supermarket chains and supporter of the development of the retail organic market in South Africa, on the performance and trends of fresh organic produce showed that fresh produce completely dominated the sales.

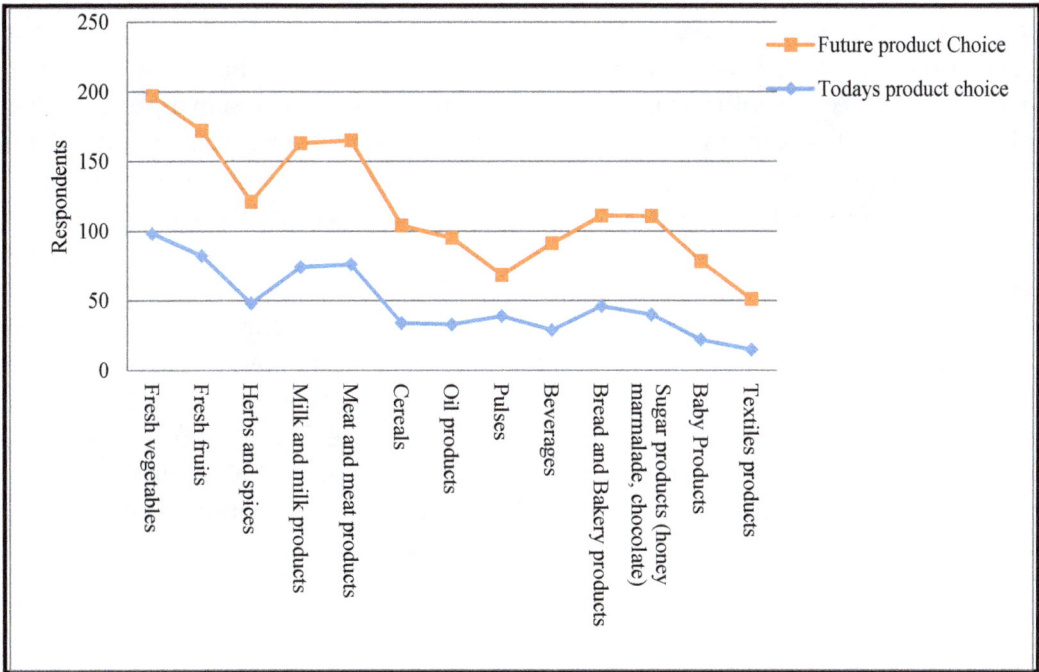

Figure 5. Demand difference between organic products today and in the future in OR Tambo District Municipality

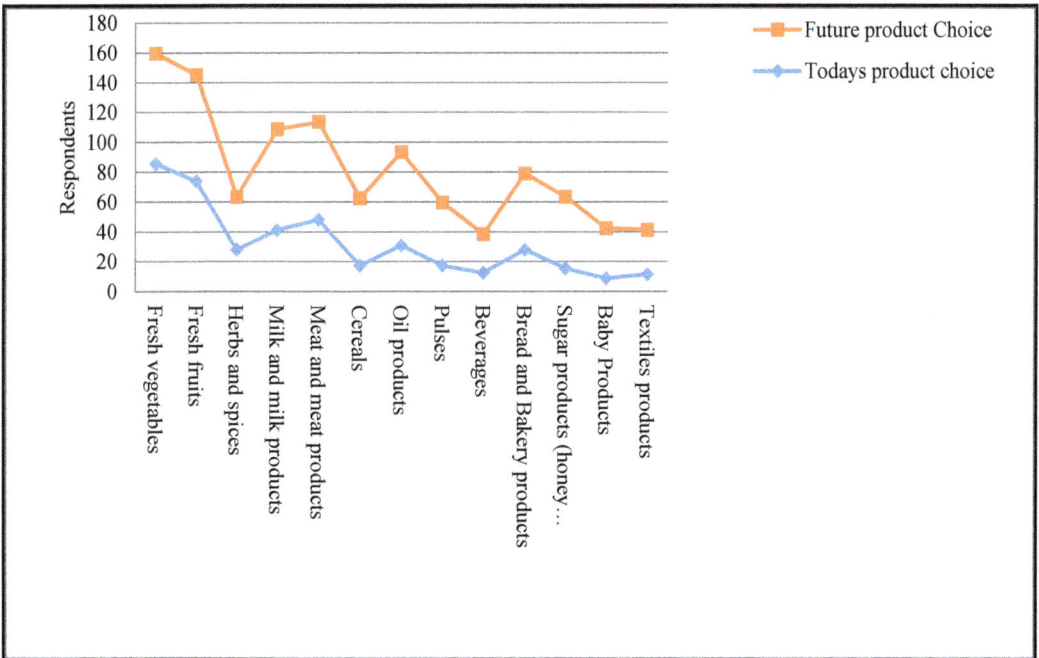

Figure 6. Demand difference between organic products today and in the future in Amatole District Municipality

This is an indication that the consumption of organic products is closely related to consumer awareness and knowledge of organic products. Increasing awareness about organic products to consumers is important to spur its demand. Most of the consumers had consumed organic products in South Africa with non consumers showing a general interest in organic products. Authors [53] state that consumer awareness of organic foods is the first step in developing demand for organic products. Yet, awareness does not necessarily equate with consumption. While organic refers to the way agricultural products are grown and processed [54], interest in consuming organic products may relate to food safety concerns where organic products may be a partial answer to recent food scares associated with production and handling (e. g. BSE, dioxins, Salmonella, etc.). Food safety issues have driven consumers to search for safer foods whose qualities and attributes are guaranteed [55]. The main reasons advanced for the consumption of organic products are that organics are healthy and nutritious, have a better appearance and taste, are affordable and are safe to consume. Identified hindrances to the consumption of organics are that they are expensive and not readily available. Price and affordability of organic products was ranked as the most important consideration among all consumers when buying organic products in South Africa.

5. Conclusions

The global markets for organic products have grown rapidly over the past two decades [8]. Currently 32. 2 million ha are being managed organically worldwide by more than 1. 2 million producers [11]. In Africa, South Africa has the third largest area (50, 000ha) under organic farming [11]. Organic production is particularly well-suited for smallholder farmers, who comprise the majority of the world's poor. The promotion of organic agriculture does not only constitute an important option for producersbut also responds to consumers' desire for higher food quality and food production methods that are less damaging to the environment. The consumers' concerns for food safety, quality and nutrition are increasingly becoming important across the world, which has provided growing opportunities for organic foods in recent years. Expectedly, the demand for organic food is steadily increasing in the developing countries. The untapped potential markets for organic foods in the countries like South Africa need to be realised with organised interventions on various fronts, which require a better understanding of the consumers' preference for organic food. Therefore, an analysis of consumer's awareness of various aspects of organic products may be considered as important ground to build the markets for organic food in the initial phase of market development. Recent analysis [53] indicate that consumer awareness of organic foods is the first step in developing demand for organic products. By identifying independent variables that explain the adoption of organic farming, the present study sought to contribute to policy formulation to promote adoption in South Africa and the rest of Africa. The identified sources of risk faced by smallholder farmers provide useful insights for policy makers, advisers, developers and sellers of risk management strategies. This information can yield substantial payouts in terms of the development of quality farm management and education programs as well as the design of more effective government policies.

Author details

Maggie Kisaka-Lwayo and Ajuruchukwu Obi

*Address all correspondence to: aobi@ufh.ac.za

Department of Agricultural Economics & Extension, University of Fort Hare, Alice, South Africa

References

[1] Tilman D, Cassman KG, Matson PA, Naylor R, Polasky S. Agricultural Sustainability and Intensive Production Practices. Nature 2002; (418)671-677.

[2] Food and Agricultural organization. FAO. Assessment of the World Food Security Situation, Report CFS2007/2. Rome; 2007.

[3] Willer H, Rohwedder M. , Wynen E. Organic Agriculture Worldwide: Current Statistics. In Willer H. , Kilcher L. (ed.): The World of Organic Agriculture. Statistics and Emerging Trends 2009. ITC Geneva; 2009.

[4] Oelofse M, Høgh-Jensen H, de Abreu LS, de Almeida GF, Sultan T, Yu Hui Q, de-Neergaard A. Certified Organic Agriculture in China and Brazil: Market Accessibility and Outcomes Following Adoption. Ecological Economics 2010; (69) 1785-1793.

[5] Byerlee D. , Alex G. Organic farming: a contribution to sustainable poverty alleviation in developing countries? Forum for Environment and Development, Bonn. Report; 2005.

[6] UNEP-UNCTAD. Sector Background Note – Organic Agriculture. http: //www. unep-unctad. org/cbtf/events/geneva5/Word Backgroundnoteorganic agriculture_ 01102007. pdf (accessed 15 June 2010).

[7] UNCTAD. Organic Agriculture: A Trade and Sustainable Development Opportunity for Developing Countries. In UNCTAD/ DITC/TED/2005/12. (ed.)UNCTAD Trade and Environment Review 2006. United Nations New York and Geneva; 2006.

[8] Sahota A. The Global Market for Organic Food and Drink. In Willer H. , Kilcher L. (Eds.) The World of Organic Agriculture. Statistics and Emerging Trends 2011. IFOAM Bonn and FiBL Frick; 2011.

[9] Willer H. , Yussefi M. The World of Organic Agriculture 2005. Statistics and Emerging Trends. IFOAM Publication, Tholey-Theley, Germany; 2006

[10] Parrott N, Ssekyewa C, Makunike C. , Ntambi S M. Organic Farming in Africa. In Willer/Yussefi (eds.) The World of Organic Agriculture. Statistics and Emerging Trends 2006. IFOAM, Bonn; 2006.

[11] Willer H. , Klicher L. The World of Organic Agriculture. Statistics and Emerging Trends 2009. IFOAM, Bonn, Germany and Research Institute of Organic Agriculture, FIBL, Frick, Switzerland; 2009.

[12] Walaga C. Organic Agriculture in the Continents. In Yusseffi M. , Willer H. (ed.) The World of Organic Agriculture: Statistics and Future Prospects; 2003.

[13] Ssekyewa C. Organic agriculture research in Uganda. Paper presented at the International Society for Organic Agriculture Research (ISOFAR) Scientific Conference, 20th-23rd September 2005, Adelaide, Australia; 2005

[14] FiBL/IFOAM: Key Results from Survey on Organic Agriculture Worldwide. www. fibl. org (accessed 9 September 2009).

[15] Moffet J. Principles of organic farming. Paper contributed at the 1st short course in organic farming, 25th-27th October, Spier Institute and Ellensburg Agricultural College; 2001

[16] GROLINK. Feasibility study for the establishment of certification bodies for organic agriculture in Eastern and Southern Africa, Report commissioned by Sida/INEC, Höje; 2002

[17] International Trade Centre. ITC. Organic Food and Beverages: World Supply and Major European Markets. www. intracen. org/mds/sectors/organic/welcome. htm (accessed 10 January 2010).

[18] ScialabbaN. E. , Hattam C. Organic Agriculture, Environment and Food Security. http: //www. fao. org/docrep/005/y4137e/ y4137e00. HTM (accessed 2 July 2011).

[19] Arnold G. The Resources of the Third World. Cassell, London; 1997.

[20] Millstone E. , Lang T. The Atlas of Food: Who Eats What, Where and Why?Earth Scan, London; 2002

[21] Mahlanza B, Mendes E, Vink N. Comparative Advantage of Organic Wheat Production in the Western Cape. Agrekon 2003; (42) 144-162.

[22] Troskie D P. Factors influencing organic production: An economic perspective. Paper presented at 12 Annual Interdisciplinary Symposium, 12 September, ARC-PPRI, Stellenbosch; 2001

[23] International Federation of Organic Agriculture Movements. IFOAM. IFOAM's Position on Smallholder Group Certification for Organic Production and Processing. http: //www. ifoam. org/press/positions/Small_holder_group_certification. html(accessed 17 April 2011).

[24] Naegeli F, Torrico JC. The Potential of Organic Agriculture to Improve Food Security. CienciAgro 2009; (1) 144-151.

[25] Zundel C. , Kilcher L. Organic agriculture and food availability. Issues Paper, International Conference on Organic Agriculture and Food Security, 3-5 May, Rome; 2007

[26] Connor D J. Organic Agriculture Cannot Feed the World. Field Crops Research 2008 (106)187-190.

[27] Badgley C, Moghtader J, Quintero E, Zakem E, Chappell MJ, Avile´s-Va´zquez K, Samulon A, Perfecto I. Organic Agriculture and the Global Food Supply. Renewable Agricultural Food Systems 2007 (22): 86–108.

[28] Statistics South Africa. STATSSA. Mid-year Population Estimates: Statistical Release P0302. http: //www. statssa. gov. za/ publications/P0302/P03022010. pdf(accessed 12 July 2011).

[29] OR Tambo District Municipality Integrated Development Plan. ORTIDP. OR Tambo District Municipality Integrated Development Plan. http: //www. ortambodm. org. za/files/PDF/pplan2012to2013. pdf (accessed 5 October 2011).

[30] Community Survey. CS. Statistical Release P0301. 1. Basic Results for Municipalities. http: //www. statssa. gov. za/publications/p 03011/p030112007. pdf (accessed 5 October 2011).

[31] Binswanger HP. Attitudes towards Risk: Experimental Measurement in Rural India. American Journal of Agricultural Economics 1980(62)395-407.

[32] Young DL. Risk Preferences of Agricultural Producers: Their Use in Extension and Research. American Journal ofAgricultural Economics 1979 (61) 1063-70.

[33] Verbeek M. A Guide to Modern Econometrics. John Wiley and Sons Ltd; 2008

[34] Greene W H. Econometric Analysis. 5th ed. Prentice Hall, New Jersey; 2003

[35] Llewelyn RV, Williams JR. Nonparametric Analysis of Technical, Pure Technical and Scale Efficiencies for Food Crop Production in East Java, Indonesia. Agricultural Economics 1996 (15) 113-126.

[36] Bradshaw B. Questioning Crop Diversification as Response to Agricultural Deregulation in Saskatchewan, Canada. Journal of Rural Studies 2004 (20)35-48.

[37] Feng XHE. , Chenqi D. Adoption and diffusion of sustainable agricultural technology: an econometric analysis. Report. Chongqing Institute of TechnologyChina; 2010

[38] Kouamé EB H. Risk, risk aversion and choice of risk management strategies by cocoa farmers in Western Cote d'ivoire. University of Cocody-AERC Collaborative PhD Programme. Abidjan-Côte d'Ivoire; 2010

[39] Yesuf M. Risk aversion in low income countries: experimental evidence from Ethiopia. Discussion paper 715. IFPRI, Washington D. C; (2007).

[40] Wik M, Holden ST. , Taylor E. Risk, market imperfections and peasant adaption: evidence from Northern Zambia. Discussion paperno. 28/1998, Department of Economics and Social Sciences, Agricultural University of Norway; 1998

[41] Sillers DA. Measuring risk preference of rice farmers in Nueva Ecija, Philippines: An experimental approach. PhD dissertation, Yale UniversityU. S. A.; 1980.

[42] Lwayo M. Risks preference and consumption decisions in organic production: The case of KwaZulu-Natal and Eastern Cape Provinces. PhD thesis, University of Fort Hare, South Africa; 2012

[43] Rahman S. Whether Crop Diversification is a Desired Strategy for Agricultural Growth in Bangladesh?Food Policy 2009 (34)340-349.

[44] Feigenbaum J. Precautionary Saving or Denied Dis-saving. Economic Modelling 2011 (28)1559-1572,

[45] Cunha F, Heckman J, Navarro S. Separating Uncertainty from Heterogeneity in Life Cycle Earnings. Oxford Economic Papers 2005(57) 191-261.

[46] Mutlu N. Consumer attitude and behaviour towards organic food: Cross-cultural study of Turkey and Germany. Master's thesis. University of Hohenheim Germany; 2007.

[47] PROVIDE. A profile of the Eastern Cape Province: Demographics, poverty, income, inequality and unemployment from 2000 till 2007. Background paper series: 1(2). Ellensburg. Cape Town; 2009

[48] Zanoli R. The European consumer and organic food. School of Management and Business, Wales; 2004

[49] Botha L. , Van Schalkwyk H D. Concentration in the South African food retailing industry: proceedings of the 44th annual AEASA conference, September 20 – 22, Grahamstown, South Africa; 2006

[50] Irwin BL. Small-scale previously disadvantaged producers in the South African organic market: Adoption model and institutional approach. Masters' dissertation. Michigan State University, U. S. A; 2002.

[51] Food Safety Agency. FSA: ConsumerAttitudes. www. food. gov. uk/multimedia / pdfs /casuk05. pdf. (accessed 5 October 2011)

[52] Vermeulen H. , Bienabe E. What about the food 'quality turn' in South Africa? Focus on the organic movement development. paper presented at the 105th EAAE Seminar International Marketing and International Trade of Quality Food Products, March 8-10, Bologna, Italy; 2007.

[53] Briz T, Ward RW. Consumer Awareness of Organic Products in Spain: An Application of Multinomial Logit Models. Food Policy2009 (34) 295-304.

[54] Organic Trade Association. OTA. Manufacturer Survey. www. ota. com(accessed 10 October 2011).

[55] ZalewskiRI. , Skawi_ska E. Food safety: commodity science point of view. Paper presentation at the International Association of Agricultural Economists Conference, August 12-18, Gold Coast, Australia; 2006

8

Organic Production of Cash Cereals and Pseudocereals

Franc Bavec

1. Introduction

1.1. Organic production of major cereal crops and maize

1.1.1. General characteristic

Monocotyledonous maize (*Zea mays* L.) and spending winter or spring sown cereals such as common wheat (*Triticum aestivum* L.), barley (*Hordeum vulgare* L.), rye (*Secale cereale* L.), oat (*Avena sativa* L.), and triticale (*xTriticosecale* Wittm. & Camus) provide the main staple foods for the world's human and animal populations. They are conventionally grown on approximately 30 % of the total world arable land, and about 50% of cereals (data from FAO for 2012), provide almost one half of total nutrition energy. Over 500 important industrial products and byproducts may be obtained from maize, and also the similar amount may be obtained from cereals. Their excellent adaptability and wide distribution belongs to the spread family *Poaceae*. For example, wheat is grown from 40° s. l. (Southern latitude) to 60° n. l. (Northern latitude); barley even to 70° n. l. Rye is less sensitive to climate changes than wheat and barley and could be sown at altitudes up to 1200 m under European conditions. Similarly, maize is grown from 53° (optimum 45°) s.l. to 35° n.l., especially wide-spread in areas with a (sub) tropical climate. Especially new crops in the region, like sweet maize in temperate climate needs to be studied carefully [1].

The aim of this chapter was to analyse data mainly from Web Sci and Sci Direct and integrate actual organic cultivation knowledge as a suggestions for further development of maize and cereal production.

The use of organic farming technologies has certain advantages in some situations and for certain crops such as maize [2], but like in case of the yield performances of 24 organic wheat yields were low and variable 3.5 +/- 1.4 t ha⁻¹ [3]. The proportion of cereals incl. maize produced organically is relatively high compared to vegetable and fruits (FAO, 2002), due to their less efficient production and more pretentious marketing. In the USA, 47% of the certified organic area was devoted to field and fodder crops in 2003, compared to 34% to pastures and 8% to fruit and vegetable crops (Economic Research Service, USDA).

In many cases farmers introduce into organic farming the old varieties and land race populations of cereals, because of their better resistance on plant diseases and due to their low fertilizer inputs, but also the yields are lower. Instead of common wheat often grown are spelt, emmer and einkorn in Austria, Switzerland and Slovenia. In spite of high inputs and low resistance on plant diseases, the new (modern) cultivars have changed many of their characteristics (grain size and leaf area have increased almost twenty times, ageing of top leaves has become slower and the period of assimilate accumulation in grains has been extended, better expressed productive tillering by cereals, cereals were bred to produce short-stem varieties to prevent lodging, …).

The competitive strength to weeds is highest for rye, less so for winter barley, oat and triticale, while wheat and summer barley are the least competitive species, due to their height of steams. Monocotyledonous weeds like couch grass (*Elytrigia repens* L.), wild oat (*Avena fatua* L.), hair grass (*Apera spica – venti* L.) and some dicotyledonous weeds like catch weed (*Galium aparine* L.) and perennial weeds like field bind weed (*Convolvulus arvensis* L.), bull thistle (*Cirsium arvense* L.), etc. are associated in cereals. Maize is associated with other group of wide-leaved – to the high temperatures sensitive annual weeds like amaranths (*Amaranthus* sp.), white goose foot (*Cheopodium album* L.), and perennial weeds. Weeds in cereals result in difficult harvesting and cause an intensive spreading of weed seeds and decrease the yield; and additional influence of weeds in maize is suppression of growth in the first leaf stages (Maier code 21-26, [4]).

In organic farming, especially early in the spring at the tillering stage (EC 22-23, see growth stages of cereals described by Zadoks [5] of winter wheat and barley, lack of available nitrogen in the soil is problematic, due to nitrogen leaching and lack of mineralization caused by low temperatures. Rye and oat are less demanding, and thus more suitable for organic production. Also in case of maize the lack of nitrogen in the spring time is a usual problem.

1.1.2. Position in a crop rotation

Success of cereal crop production primarily depends on the choice of the previous crop, especially in less fertile soil. In conventional farming systems, there is a possibility to compensate for a less suitable foregoing crop by using higher rates of fertilizers and pesticides, but in organic farms, residual soil fertility from crop residues primarily determines the potential yield of a cereal crop. Rotation is also essential to prevent the build-up of pests and diseases; in particular foot and root rot diseases.

In mixed farming systems (incl. animal and crop production) where a grass/clover ley is part of the rotation, cereals can be grown after incorporation of the grass/clover mixture or after a more N-demanding intermediate crop grown after grass-clover incorporation, such as potatoes. In stockless farming systems (without livestock), legumes, cereal-legume mixtures, root crops, or oil seed crops are suitable foregoing crops, provided sufficient N is left from the root- or oil seed crops. In exceptional cases, cereals can be grown for two consecutive years, for example a winter and summer cereal. In that case, it is best to use a less demanding crop (rye or oat in cool climates, maize or millet in warmer climates) as the second crop. Compared to conventional rotations, organic crop rotations generally have a lower proportion of cereals, not exceeding 50%. Above this rate, only crops with low nutrient demands (rye, oat, triticale) can be used. If a legume such as red clover or alfalfa is used as foregoing crop, phytotoxicity od the decomposing residues may be a problem for the germinating seeds of cereal crops [6]. Therefore these foregoing crops should be ploughed under at least three weeks before the cereal is sown.

Because the maize is more demanding for N than temperate cereals, maize is usually grown after a leguminous green manure or fodder crop [7]. Maize is often only rotated with soybeans in conventional production systems, but a corn-soybean rotation does not provide adequate N in an organic system, as most of the N is removed with the soybean. Thus, organic rotations need to include additional leguminous crops besides soybeans or use additional animal manure. A common organic rotation used in the Midwestern USA consists of maize, soybeans, a cereal, legume green manure, and a fifth crop that is varied from year to year [8]. The effects of 2- versus 4-year rotations on corn and soybean yields were studied in organic and conventional systems in Minnesota [9]. Maize in the organic four-year maize-soybean-oat/alfalfa-alfalfa rotation yielded only 7-9% less than in the conventional two-year rotation, and almost twice the yield in the organic 2-year rotation. In another study in Iowa, yields of organic and conventional corn were equivalent in a similar 4-year rotation [10].

Cereal crops also have a useful function for the following crops in an organic rotation, due to their fibrous root systems that may loosen deeper soil layers, depending on the particular crop (Table 1). Moreover, monocotyledonous crops generally are not susceptible to diseases of dicotyledonous crops and, thus, provide a break in the build-up of diseases of dicotyledonous crops.

| Crop | Root | | Tolerance to yourself | Foregoing crop quality (generally) | Land coverage | Pertinence like cover crop |
	Amount	Depth				
Winter wheat	+	++	+	+	+	+++
Spring wheat	+	+	+	-	+	++
Winter rye	++	++	++	++	++	++
Winter barley	+	+	+	+++	+	+++
Spring barley	+	+	++	++	+	++
Oat	++	++	++	-	+	++
Millet	++	+++	+	-	+	++

Explanatory notes: +++ good; ++ proper; + poor; - improper

Table 1. Properties of cereals when used as a foregoing crop (arranged according to Demo, Bielek *et al.* [11].

Crop rotation is also essential for regulation of weeds in cereal crops. Summer cereals should be preferentially grown when winter weeds, such as mouse foxtail (*Alopecurus myosuroides* Huds.), scentless mayweed (*Matricaria maritima* L.), catch weed (*Galium aparine* L.), hair grass (*Apera spica – venti* L.), field poppy (*Papaver rhoeas* L.) are problematic, while winter crops can better be grown when summer weeds (such as wild oats or wild rape seed) occur more frequently. Cereal crops show different competitive strengths against weeds. In general, winter cereals are more competitive than summer cereals, rye more than wheat, and oat more than summer barley. In the low-stem cereals under sown cover crops (from genus *Trifolium*) can be included, because they compete generally well with weeds after harvest and represents the further leguminous in the crop rotation.

Crop rotation also contributes to the regulation of pests and diseases. Black stalk or take-all disease (*Gaeumannomyces graminis*) and eyespot disease (*Pseudocercosporella herpotrichoides*) represent major fungal diseases of cereal crops that are strongly affected by rotation. Take-all is commonly suppressed by antibiotic-producing *Pseudomonas fluorescens* bacteria when wheat is grown without rotation [12], but this is not an option for organic wheat production. In organic farming systems, take-all can be reduced by using soil-improving, non-susceptible annual crops in the rotation, because the pathogens do not persist in soil for a long time period. Suitable *G. graminis*-interrupting crops are maize, oat, potatoes, sugar beet, legumes, rape, or flax. High microbial activity and diversity, promoted by a complex rotation, also contribute to take-all control [12]. Cereal crops need to be interrupted by other crops for at least 2-3 years to limit eyespot occurrence. The most suitable crops for this purpose are clover or alfalfa for 2 - 3 years. Very suitable are sequences of these crops, for example maize - oat, potatoes - oat, potatoes - legumes etc. However, to prevent the build-up of smut diseases, such as *Tilletia controversy* on wheat and *Ustilago maydis* on maize, longer rotations of 6-8 years are re-quired. This is especially true when organically produced seeds are used, as the common seed treatments are not allowed. Too many cereal crops in the rotation may also enhance the risk of foliar diseases such as powdery mildew (*Erysiphe graminis*). Due to low resistance of cereals to the diseases, one of the most effective measures for stable yielding is intercropping of different species of cereals [13] or just different varieties (with the same harvesting time) of one cereal species [14].

The most frequent cereal pests are aphides, damaging the assimilatory organs and spikes. As cereal aphids are quite specific, it is important to rotate with non-cereal crops. In Europe and Western Asia, the corn ground beetle (*Zabrus gibbus*) can be very damaging, especially on wheat [15]. The most damaging pests in maize are corn borer (*Ostrinia nubilalis* L.) and Western corn rootworm (*Diabrotica virgifera* L.), becoming new pests in Eastern Europe. Rotation of cereals with legumes and beet is a sufficient measure against these pests, which' larvae damage green grains and adults feed on crop flowers and grains during milk ripe stages. Wire worms (larvae of the click beetle) can be a major pest in organic crop production, especially after incorporation of large amounts of crop residues, for example perennial fodder crops or a cereal crop with under sown grass-clover. Populations of click beetles are generally reduced by cultivation operations. As root crops and rape seed require multiple cultivation operations, these crops are more suitable for rotation.

1.1.3. Cereal crops management practice

Variety selection

The main criteria for selection of cereal crops and varieties for organic production are: suitability for the certain location, yielding ability, stability and plasticity, large rooting and nutrients uptake ability (especially of nitrogen), resistance to fungal diseases, high competitive ability towards weeds, and stress tolerance (drought, water logging, extreme temperatures). The lower internodes should be relatively short to prevent lodging, but the upper internodes long to support higher ability of assimilation during the period of grain forming, even if the leaves are damaged by fungal diseases. Mid-height and bearded varieties seem to be optimal as these produce good grain yields with fewer productive stalks but with more grains per spike. It is important to select varieties with resistance or tolerance to fungal diseases prevailing at the particular location, as synthetic fungicides are prohibited in organic cereal production, and copper fungicides are generally not used on these crops. In case of maize the special attention must be done to the sowing appropriate FAO groups of hybrids, depending on geographical latitudes (FAO classification valid for Corn Belt in USA, number of growing days for each group growth out of this latitude are longer) and other production demands, and also to the target type of grains (dent, flint, semi dent or flint, pop, sweet) due to it's potential use.

Older varieties of cereals usually have favourable features to low pretentiousness of production inputs, with possible respect to grain quality, but lower productivity. Such varieties can be used for particular contracts with processors, where the price of a specific product balances lower yields.

Preparatory soil cultivation

Soil cultivation is primarily needed for destruction of crop debris and stubble of a previous crop, weed control and mineralization of nutrients, but also to enhance porosity for aeration and water and root penetration. Several cultivation operations can be distinguished, each with their own objectives, for example: ploughing to turn the soil so that crop residues, weed seeds and plant pathogens are buried below the immediate rooting zone, tillage to loosen the soil without turning it, harrowing to prepare a loose seedbed (4-5 cm deep), and hoeing to remove weeds.

The timing of cultivation is extremely important, because the wrong timing can destroy the soil structure, and nullify the benefits of proper organic matter management. This was shown for a biodynamic farm, where soil had been ploughed in a too wet condition, leading to soil compaction. The soil structure was not significantly better than that of a neighbouring conventional farm in comparison to the much better soil structure of permanent grassland. For winter cereals, the soil is obviously prepared for seeding in the fall, but for summer cereals and maize there is a choice of fall or spring soil preparation, depending on the climate. When springs are generally wet, for example in land climates with heavy snow fall, fall ploughing may be the only option, even for spring crops.

The depth of cultivation is another important choice. Common ploughing should not be deeper than usuall, because of live and well structured layer of the soil (suggested 20-25 cm). Even higher ploughing must be step-by step i.e. cm by cm. Recent research has shown that soil organic carbon can be preserved by minimum tillage (shallow tillage, 10-15 cm deep). This also results in minimal disturbance of soil life, particularly fauna and fungi. Many organic farmers opt for shallow tillage, using the so-called eco-plough. After potato for winter cereals or annual cover crops for maize, direct sowing is possible also in organic production. This has resulted in considerable reduction in erosion and nutrient losses, and in improvement of soil structure. However, no-till is hardly an option in organic farming, as herbicides cannot be used, and hand-weeding and mechanical in-row hoeing are not feasible for field crops. Minimum tillage may also result in an increased occurrence of weeds, especially hair grass (*Apera spica – venti* L.), camomile (*Chamomilla*), couch-grass (*Elytrigia repens* L.), creeping thistle (*Cirsium arvense* L.), barnyard grass (*Echinochloa crus – galli* L.), and pig-weed (*Amaranthus* L.), which thrive on compacted soils.

The number of operations and their intervals ultimately determine weed infestation rates. To avoid weed problems in the beginning of crop growth, soils are sometimes harrowed an extra time before final seedbed preparation, so that weeds germinate, start to grow, but are killed by the second harrowing. The first operation is called preparation of a 'false seedbed'. A certain level of weed cover could be tolerated when crops are past their most sensitive stage [16], but noxious weeds should be removed before setting seed. Too frequent tillage operations lead to soil erosion and nutrient and carbon losses, and ultimately to poor soil structure.

Nutrition of cereals and maize

The uptake of nutrients with aboveground maize yields can reach up to 278 kg N ha^{-1}, 46 kg P ha^{-1} and 171 kg K ha^{-1} [17], in cereals approximately up to half of the amount. Target fertilization rate depends on available nutrients in the soil and target yield. Target fertilization rate depends also on climatic conditions, genotype, and production system like plant populations [18, 19]. According to the EU regulations limited input with organic fertilizers is 170 kg N ha^{-1} as an average farm^{-1}. It means that is impossible input the same levels of nutrients in organic production comparing with conventional agriculture. Deria et al. [20] considered that several factors causes grain yield differences (organically grown wheat was: increased at one site by 17%; decreased at three sites by an average of 27%, and not changed at another three sites) between the organic and conventional wheat: like lower nitrogen supply, and lower extractable-P of the organic wheat.

Cereals absorb nutrients mainly from the upper soil layer. Their development depends on available nutrients, especially phosphorus and nitrogen, released from organic fertilizers or crop residues by mineralization.

Nitrogen nutrition is often not limited by the total nitrogen in soil, but by a deficiency in certain growth stages, especially at the tillering stage and generative stages of winter cereals. Release of mineral nitrogen is often maximal in mid summer, often resulting in too high nitrogen levels in brewing barley, but still fairly low nitrogen (and gluten) levels in bread wheat.

Besides crop residues, organic cereal crops may be fertilized with animal manure, especially on poor soils, at rates of 4-10 t ha^{-1} (up to content of 180 kg N ha^{-1} year^{-1}). Also legumes as preceding crops improve nitrogen nutrition on subsequent winter wheat, e. g. the Nitrogen Nutrition Index at flowering (NNIf) = 0.51 +/- 0.12 for a crop rotation with a rate of legumes over 37% vs. 0.41 +/- 0.11 for a crop rotation with a rate of legumes under 25% [3]. In the case of comparing conventional animal production, organic animal production, and organic cereal production residual fertility effects from the preceding crop produced higher yields in organic animal production winter and spring wheat than in conventional. Cultivation of winter wheat in organic animal production was a more efficient use of nitrogen resources than conventional [21]. The findings that microbial properties and N availability with 182-285% increase in potentially mineralizable N, for thus the yields of crops differed under different organic input (regimes cotton gin trash, animal manure and rye/vetch green manure), [22].

Organic fertilizers may also be combined as phosphorus and potassium fertilizers which are allowed (depending on the certification agency).

An additional nutrient dressing can be applied at the generative stages (EC 20–35) by dispersion of dung (5-7 t ha^{-1}) or slurry (20-40 m^3 ha^{-1}). Higher rates of dung or slurry can cause expansion of ruderal weeds (docks, goosefoot, hardship, scentless may-weed, barnyard grass, etc.). Harmonious fertilization maintains the competitive strength of crops, promoting fast foliage growth and better soil coverage.

Additional supply of nitrogen is possible by intercropped legumes. The data showed that nitrogen content of the wheat grain and whole plant biomass significantly increased when the density of beans in the intercrops increased; this was reflected in a significant increase in grain protein at harvest of organically grown wheat [23]. The clear and significant differences in wheat yield and protein content between the organic and conventional systems suggest a limited supply of available N in the organic fertility management system which is also supported by the significant interaction effect of the preceding crop on protein content [24].

Seed selection and sowing

According to EU regulations, if available, certified organically grown seed is required for certified organic crop production (EU Regulation No. 834/2007). As sufficient certified organic seeds of cereal crops and maize are on the market, European organic growers are bound to use these seeds. The situation is different in North America and Canada, where grower-saved seeds are often used, or commercial, conventionally grown seeds without fungicide coating.

The selected sowing rate should ensure an optimum stand density that results in soil shading and weed suppression. Sowing at high density at a narrow row distance is more effective to control weeds in a cereal crop than wide row spacing combined with mechanical weed control. Early sowing is not as effective for weed control as sowing after repeated harrowing (preparing a false seedbed to kill germinated weeds). Finally, the sowing depth is crucial: not too deep to avoid root and foot diseases, and not too shallow to avoid drying of the seed.

Both, cereals and maize can be intercropped by legumes; in the case of maize intercropping with climbing bean is suggested and in case of cereals with pea and beans. Sowing of obove mentioned intercropped crops is separate, depending on suitable periods for seed emergence and growth.

Weed control

High weed pressure is one of the main problems of organic cereal and maize production. Weed seedbank size was larger under the organic system of maize production than conventional system; and for example differences among maize management systems depends mainly upon weed control efficacy rather than upon tillage effects [25].

The most problematic weed species in cereal production are couch-grass and creeping thistle. Application of herbicides is impossible and thus different ways of weed regulation must be used. The aim of such regulation is not to completely destroy weeds, but to keep their coverage under the damage threshold. The emphasis is put on prevention of introduction and spread of regenerative organs of weeds (seeds, stolons, etc.), and on indirect control like promoting optimal growing conditions for the crops that enhance their competitive strength against weeds. Among the preventive measures belong selection of a location that is suited to a crop's demands, a varied and well-balanced crop rotation including fodder crops, proper soil cultivation, use of barnyard manure, harmonic fertilization, proper choice of species and varieties, proper sowing, prevention of weed introductions, optimum time and way of harvesting as well as post-harvest operations. Finally, care must be taken that noxious weeds don't spread from field margins and surrounding natural vegetation.

If the preventive measures are not effective enough, direct weed control treatments must be applied: mechanical, physical and biological methods. Harrowing using different harrow types (weeders, net harrow, and spike harrow) can be used for deep-rooted crops (maize, sorghum) before the plants sprout but after germination of weeds. A second harrowing can be applied shortly after the stage of 2-3 true leaves (Maier code 21-23). Line weeding (with brush cultivators or deer-tongue cultivators) can be applied in wide-row cereal crops on heavy soils (row spacing over 15 cm) or to combat overgrown weeds when cultivation was delayed, for example due to wet conditions. Thermal weed control (using propane-butane burners) can be used to eliminate all weeds before sprouting of the crop, or to put down dicotyledonous weeds in wide-row crops like maize. Biological weed control by plant pathogens or insect pests is still hardly used in organic agriculture.

Harvesting, post-harvest treatments and storage

Organic grains need to be harvested and stored separately from conventional grains. Perfect threshing and pre-cleaning of the grain is a basic expectation for proper harvesting of cereals. Grain needs to be dried to 13.5 - 14% moisture. Drying of grain is performed gradually, removing only about 2% water at a time if it originally had moisture content over 20%. Organically grown grain should be dried like seed material, preventing too high temperatures, because sprouted grains can be used for human consumption.

Grains of cereals and maize can be infected by fungi that produce mycotoxins (primarily *Fusarium* species), which can seriously affect human and animal health. Therefore mouldy grain should not be used, not even for feeding purposes. To prevent mould infection, grain temperature, moisture, smell and pest invasion must be monitored, and the storage silos should be aerated. The air temperature should be 5 °C lower than the grain temperature. The optimum temperature for grain storage is 5 – 10 °C, and the temperature should not exceed 20 °C. Storage pests, especially grain weevils (*Sitophilus granarius*), grain moths (*Nemapogon granellus*), and flour mites (*Acari*) can be a serious problem for organic grain, as they can only be controlled within the limits given by regulations. The storage silos need to be cleaned before use. Natural pyrethrums can be used to control invading pests, but these substances have a short-term effect only (5-6 hours), so the application has to be done repeatedly. Natural pyrethrums do not affect warm-blooded animals. If these disinfection procedures are followed, limited occurrence of pests can be expected.

1.1.4. Labour use and economics

Costs and prices

The yields of organic cereals and maize are usually somewhat lower than those achieved under conventional management, mainly due to limited nitrogen availability. However, in a long-term experiment with a five year rotation, yields of organic maize were very similar to those of conventional: 6.4 versus 6.5 tons/ha [26]. In drought years, organic maize yielded even more (28-34%) than conventional maize, thanks to the greater water-holding capacity of the organically managed soil. But in drought conditions in case of organic maize yielded just 38% and 137% relative to conventional were formed [27].

The direct costs for organic grain production are frequently lower than those for conventional production, as there are no expenditures for pesticides, growth regulators (for small grain cereals only) and synthetic fertilizers. These lower costs may sometimes compensate for the lower yields, even at conventional market prices, as observed for maize production in the Corn Belt in the USA [28]. However, for calculation of net income from a crop, opportunity costs need to be included, as less profitable leguminous rotation crops are needed to provide sufficient nitrogen for a corn crop [26, 29]. In addition, more expensive seed, more frequent cultivation and the associated fuel costs, and higher costs for manual labor for preparation of natural fertilizers and weed control detract from the profitability of organic small grain and corn production[26]. Taking the lower yields per ha and all costs per ha into consideration (Fig. 1), the production of cereals and maize would mostly not be economic at conventional prices for the products. Premium prices are essential for a positive balance [26].

Price premiums vary considerably, from 15 to 50% for organic cereals (Sullivan, 2003), and from 20 to 50% for organic corn [29]. Even a premium price of 200% of organic over conventional maize has been reported [30]. The height of the price premium depends, of course, on the kind of cereal, the quality, the extent of processing, the kind of market, and supply versus demand. Especially for maize, specialty crops (sweet maize, pop corn, conserved grains) fetch higher prices than regular feed maize. The premiums will likely remain fairly high as long as the demand for organically grown grains is increasing at the same pace or faster compared to the increase in organic grain production.

However, the economic net return was higher for conventional farming system, but when the economic subsidies from EU were considered, the integrated net return was higher than under organic farming system for wheat and soya [31]. As the demand for organic meat is rising, this should also increase the demand for certified organic feeds, and possibly the premium prices (as long as the number of certified organic grain growers is not expanding as fast or faster). Altogether, it is difficult to predict future organic grain prices [29].

Markets

Depending on the quality of the grain, it can be used as animal feed or for human consumption; grading standards for grains for human consumption are higher than for feed grains [32]. In mixed organic farms, grains are commonly used as livestock feed on the farm itself. Fodder grains can also be sold on the regional, national or international market. However, organic feed grain is generally not transported over great distances. Although it is more difficult to grow organic corn and cereal grains for human consumption due to the higher quality standards, they result in higher net profits than feed grains. Organic cereal grains for human consumption are commonly marketed in different ways than conventional grains, since the buyers are usually smaller-scale than buyers of conventional grain [33]. Organic grain is often produced on a contract, and producers may have more grain than a typical small-scale buyer would need. Thus, organic grain often needs to be stored, and it may be beneficial for an organic grower to have his/her own storage facilities [33]. Some organic farmers process their grain into flour and bread or other cereal products, and sell it directly to consumers at home or at the local market. The greater the share of the production chain is controlled by the farmers themselves, the more profitable the production of cereals will be.

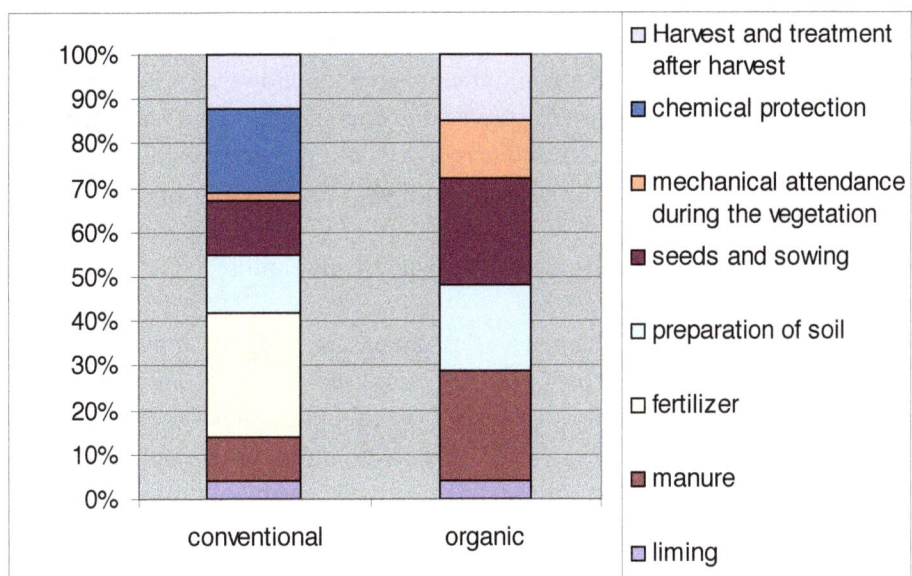

Figure 1. Comparison costs of convencional and organic farming (growing wheat) – variable cost and fixed cost of machines (Moudrý, 2005) – not published data

1.1.5. Ecological impact

The results show a markedly reduced ecological footprint of the organic systems (icluding biodynamic) in production of wheat (*Triticum aestivum* L. 'Antonius') and spelt (*Triticum spelta* L. 'Ebners rotkorn'), mainly due to the absence of external production factors. When yields were also considered, the organic systems again had a reduced overall footprint per product unit and increased ecological efficiency of production. Thus, this farming systems present viable alternatives for reducing the impact of agriculture on environmental degradation and climate change. Nevertheless, room for improvement exists in the area of machinery use in all systems studied and yield improvement in the organic farming system [33].

2. Organic production of pseudocereals

2.1. General characteristics

Pseudocereals described in this book include dicotyledonous field crops of various botanical families i.e. common buckwheat (*Fagopyrum esulentum* Moench) family *Polygonancae*, quinoa (*Chenopodium quinoa* Willd.) family *Chenopodiace* and grain amaranths (*Amaranthus caudatus, A. cruentus, A. hypohondriacus*) family *Amaranthaceae*, which were traditional food in Asia from 10th to11[th] BC, of Inca's and Aztec civilization, respectively. Their production (Table 2) has become more and more interesting, due to grains nutritional and healthy value, especially if they are gown organically. For those the own experiences at the farm and in research are again evaluated within the scope of new findings cited in research databases.

These crops have an extensive diversification among genotypes, which were utilized and adapted under specific environments; consequently their increasing introductions in the new environments differ in expression of genetic potential of yield and duration of vegetation period to reach full maturity. The main limitations for production under European temperate climate and its above sea altitudes are frost free days during vegetation period and necessary effective temperatures for active photosynthesis (assimilation), except in buckwheat where the temperatures exceeding 24°C cease pollination and stop growth due to short day-length in some genotypes. The main preferences are well adaptation of quinoa and amaranths to dry conditions, soil pH and salinity, and also C_4 pathway in amaranths.

The grains vary in morphological characteristics, shape and weight; for example 1000-g seed weight in buckwheat varies from 18 to 32 (38) g, in quinoa from 2 to 6 g, and in grain amaranths from 0.3 to 1 g. These pseudocereals have a great possibility of niches organic products with high nutritional and healthy value, suitable to diabetics and people with celiac disease, and prohibited milk proteins, etc. In comparison with cereals, grains of these pseudocereals are gluten free food, suitable for people with celiac disease, with higher protein content and their rich amino acid composition, rich in fibers, polisaturated (gamma) fatty acids incl. squalene in amaranths and minerals, etc. (Table 3), however composition depends also on milling fractions and food processing. All three pseudocereals are rich in lysine (the first limiting essential amino acid in cereals); quinoa also in histidine, isoleucine and metionine + cystine, its' consumption

	Buckwheat	Quinoa	Amaranths
Production			
Origin of utilization[a]	Asia, highlands	Andes mountains	South America
Main producers	Russia, China, USA, Japan, EU[b]	Peru, Bolivia, Argentina[c]	USA, Peru, Bolivia[b]
Yield (kg grain ha[-1]) - approximate	800-1000[b]	400-1200[b], 640-920[c]	400[b]-1500[a]
minimum, maximum	500-2200[b]	435-6591[b]	10[d]-4000, 5000[b]
Requirements			
Growing season (days)	100	120-240[e], 140[f]	105-160[b], 150[f] 90[g,h]-100[h]
Min. germination temperature (°C)	7[f]	5[f]-7[b]	12, 15[i]
Frost resistance to (°C)	-1[f], (-1.3 to -2.9)[j]	-3 do -15[k]	0
Temperature stop assimilation (°C)	10	8	15
Transpiration coefficient (kg water kg[-1] of dry plant matter)	500-600[b]	400	200-333[k]

Sources: [a] [35], [b] [36], [c][37], [d] [38], [f] [39], [g] [40], [h] [41], [i][42], [k][43], [j][44], [k][45]

Table 2. Pseudocereals production characteristics and environmental requirements

less than 50 g of grain per adults' daily need, it is also ideal reference of essential amino acid pattern for children according to FAO/WHO/UNO standards (except for valine); amaranths are a substitute for meat meals, especially due to rich content of methionine, cisteine and arginine [45, 36]. For those mixtures with cereals supports balanced meals in amino acid composition, especially for vegetarians.

	Buckwheat				Amaranth	Quinoa
	Seed	Groat[b]	Dark flour[a]	White flour[a]	Variation	Average[f] (variation)
Protein	11-15,12.3[a]	16.8	14.1	6.4	13.3[c]-17.9[d]	14.7 (9.6-22.1)
Starch	73.3[a]	67.8	68.6	79.5	62[g]	58.2 (46.9-77.4)
Fat	2.3[a]	11	3.5	1.2	5.1-7.7	7.2 (1.8-8.2)
Fibre	10.9[a]	0.6	8.3	0.5	8[g]	6.4 (1.1-13.4)
Ash	2.1[a]	2.2	1.8	0.9	3[g]	3.4 (2.4-9.7)

Sources: [a] [47], [b] [48], [c] [46], [d][49], [f] average data from Fleming and Galwey [50], [g][39].

Table 3. Nutritional composition of pseudocereals (% of dry matter)

In addition pseudocereals are characterized by many health-important substances like antioxidants – flavonoids: rutin, isovitexin, etc., flavones, phytosterols, fagopyrins and thiamine binding proteins and prebiotic activity in buckwheat. Consumption of buckwheat can improve diabetes, obesity, hypercholesterolemia, constipation as well as reduce blood pressure and cellular proliferation. The most important component in amaranths is squalene, which affected synthesis of cholesterol, possesses antisclerotic properties and may be used in the prophylaxis against cancer (see references in Bavec and Bavec[36]. From quinoa grains leached bitter saponins represent natural substances used in organic detergents, soaps, shampoos, cosmetics, etc.; the high-growth cultivars of amaranth are suitable for energetic purposes.

Among pseudocereals buckwheat is the most weed resistant crop which results from its fast development in early growth stages and might due to allelopathy [51].

In buckwheat do not exist disease and pest infestation, which might cause economically important damages, except emerging plants by nematodes. Also quinoa and amaranths should not require special pest and disease control. But several species of cosmopolitan polyphagos pests (e.g., *Agrotis ipsilon* Hufnagel, Lepidoptera: nuctuidae) may cause economic losses. Some of widespread pests endemic to the Andean region may cause up to 50% loss of yield in quinoa, especially two species of moths (*Eursacca quinoae* Povolyny, *E. melanocampta* Meyrick) [52]. Emerging plants may also be damaged by flea beatles and caterpillars. Diseases caused by *Peronospora* sp., *Sclerotinium* sp., *Phoma* sp., *Botrytis* sp, and *Pseudomonas* sp. may have an influence on yielding of quinoa. However, viruses found on spinach or beets have been observed also in quinoa fields.

2.2. Position in rotation

Because of previously mentioned plant diseases and possible polyphagous pests which could appear, and increase of weeds, each pseudocereal may not be planted again on the same field for at least 3 years; 5-6 years is suggested in organic farming. A negative consequence of incorrect buckwheat crop rotation is also increased occurrence of necrosis and the appearance of some disease signs, especially root diseases caused by *Gaeumannomyces* sp. In spite of amaranths, quinoa seeds also do not exhibit dormancy and they germinate when conditions are suitable; the plant itself though, in wild form, may remain in soil for 2 to 3 years without germinating.

According to different requirements about nitrogen needs in amaranths (high nitrogen accumulative C_4 plant), recommended rest of available nitrogen (after precrop and/or additional fertilization with organic manure) is high in comparison with buckwheat, due to lodging in the case of high value of available nitrogen in the soil [36]. However, suitable precrops crops are cereals, grain legumes and perennial legumes. Buckwheat appears as a very suitable precrop for fiber flax due to its strong competitive strength against cough-grass and dicotyledonous weeds. Considering its shorter growth season can be used as stubble crop after early harvested previous crops (early potatoes, barley, crop-legumes mixtures for green feeding). Because amaranths and quinoa growth slowly than buckwheat during early growth stages, the field in quinoa, and especially in amaranths, must be free of weed competition, for those precrop must suppressed the weeds. Buckwheat, sown as crop, cover crops or for green manure, may produce allelophatic substances that could inhibit weeds [51] and also follows crop in rotation.

2.3. Crop management

Species and varieties selection

In common buckwheat exist more than 4500 different varieties and land race populations, but in every climate special attention should be given to choosing appropriate cultivars, because of different photoperiodical reactions.

From more than 60 wild species of genus *Amaranthus*, four of them were selected like grain amaranths. They are different among each others in germination [42], growth, habitus, inflorescence, ripening and seed colour [53].

The genus *Chenopodium* consists of 150 species, classified into 16 sections. Few plants from section *Chenopodium* are used for human consumption. Quinoa is the most important species from genus *Chenopodium*. In the first stage, cultivated genotypes originated from 6 ecotypes and 4 plant populations from different regions [36]. Also in amaranths and quinoa introduction of new cultivars must based on previous analyses of trials results in every new climate.

Fertilization

Pseudocereals requirements for nutrients need vary from low to very high. Organic fertilizers that are left over from the preceding crop or given stable manure in autumn provide adequate yielding in the organic production system. Modest available level of nutrients results in low yields, except in case of some heavy metals in amaranths (important remediate plant), and Ca-bound phosphorus in buckwheat, which efficiency might be influenced by acidify of the rhizosphere [54] compared to spring wheat. In buckwheat in case of more than 20 g NO_3-N kg^{-1}dry soil to the depth of 0.3 m fertilization with nitrogen manure is not allowed, due to possible lodging problems; it should instead be used for the previous crop in the crop rotation [36]. Although amaranth yield is responsive to available nitrogen in the soil like maize, but a high level of available nitrogen can negatively affect grain harvest in terms of increased lodging, and delayed crop maturity [55]. Nitrogen application up to an Nmin target value of 140kg N ha^{-1} raised protein concentration in grain of amaranths, and maintained the content of essential amino acids in protein. Essential amino acids in grain fertilized to the target value 140 kg N ha^{-1} was higher (397 g kg^{-1}) than the standard requirement for preschool children (339 g kg^{-1}). Among essential amino acids, only valine concentration responded to nitrogen supply. Leucine was the limiting amino acid in grain protein [56, 57]. In quinoa, compensation is suggested by organic fertilizers at an application rate of about 120 kg of N ha^{-1} [58] for the expected enhancement of grain yield.

Sowing

Buckwheat is often sown as stubble crop, amaranths and quinoa are main crops. Different sowing methods, including sowing rates variy among buckwheat (mainly suggested 200 – 300; row spacing is from 12.5 to 25 cm, sowing depth is about 2-3 cm, sowing rate is 1,0 1,5 mil.

seeds, i.e. 40-60 kg per ha. Late sowing, for example on weedy fields, is recommended to be done using closer row spacing and higher sowing rates. Greenhouse experimental data in quinoa showed that optimal plant populations varies from 30 plants m^{-2} [58] to an optimal plant population of 140 plants m^{-2} [59], to 200 plants m^{-2} in the 2.3 field trials (Jacobsen et al., 2005) with interrow spacing 0.4 to 0.8 m. In amaranths 0.12 to 08 m interrow spacing, 150 to 200 seeds m^{-2} and final plant population about 40 plants m^{-2}. From variable data we can stress that in *A. cruentus* cultivars would be recommended above 30-cm row spacing and populations of 74000 and 272000 plants ha^{-1} [40].

Yields of *A. cruentus* cv. 'G6' were affected by growing season and date of sowing. There was a higher protein content (165 g kg^{-1}) and lower (39.3 g kg^{-1}) in grain of plants sown in May and June, respectively [57].

Harvesting, post-harvest treatments and stocking

According to botanical distinction harvesting, postharvest techology and food processing [60] vary in comparison with cereals, except seeding machines, harvesters and mills in buckwheat. Buckwheat is harvested when about 75% of the seeds have reached ripeness, in quinoa when at the maturity stage, plants become lighter and more yellow, and the leaves fall off and in amaranths the appropriate harvesting period is indicated by the yellowing of bottom leaves and dry seed [36]. After harvesting, which can be done with special adaptation of cereals or clover threshers, must follow immediate winnowing, after-drying to 10-12% water content and placing in a slot-floor storages equipped with an air ventilation system. Storage facilities must be dry and airy.

2.4. Labour use and economics

Pseudocereals as low input crops (labor, fertilizers) have a petty differences between yield in conventional and organic farming system, where production should be more profitable according to higher prices of organic products.

3. Conclusions

Cereals and maize are the most important food and fodder crops also in organic farming, they are essential rotation crops, as long as they are not taking up more than 50% of the crops in the rotation. Cereals are more suitable for cool and humid climates than maize, due to the slow initial growth caused by low temperatures and less expressed weed competition. Well structured soils and enough available nutrients in the soil based on crop rotation and inputs of organic manures, and systematic use of other preventive and corrective measures for pest, disease, and weed control will contribute to stabilization of yield at about 80-90% in comparison with conventional crops, depending on climate and soil type. Using organic price premiums for organic crops, net incomes from organic cereals and maize are higher than those of their conventional counterparts.

Buckwheat, grain amaranths and quinoa are very suitable crops for organic farming, especially for small scale farms. They extend agro-biodiversity and range of niches organic products, also for special uses. For an effective production selection of varieties and appropriate growing conditions should be studied and knowledge of post harvest technology and food processing must be detailed and up to dated.

Author details

Franc Bavec*

Address all correspondence to: franci.bavec@um.si

University of Maribor, Faculty of Agriculture and Life Sciences, Hoce, Maribor, Slovenia

References

[1] Fekonja M, Bavec F, Grobelnik-Mlakar S, Bavec F. Growth Performance of Sweet Maize under Non-Typical Maize Growing Conditions. Biol. Agric. Hortic. 2011;(27) 147-164.

[2] Pimentel D. Economics and energetics of organic and conventional farming. J. Agric. Environ. Ethics, 1993;63 53-60.

[3] David C, Jeuffroy MH, Henning J, Meynard JM. Yield variation in organic winter wheat: a diagnostic study in the Southeast of France. Agron. Sust. Develop., 2005; (25) 213-223

[4] Shütte F., Maier U. Entwicklungstadien des Mais zum Gebrauch fuer das Versuch-wesen, die Beratung und die Praxis in der Landwirtshaft. Biol. Bundesanst. F. Land-u Fortwirtsch., Merkblat. 1981;27 p.

[5] Zadoks J. C., Chang T. T., Konzak C. F. A decimal code for growth stages of cereals. Weed Res., 1974;14 415-421.

[6] Urban J., Šarapatka B. Ekologicke zemedelstvi (Organic farming)-ucebnice pro skoly i praxi. I. Dil, MZP Praha, 2003;280 p.

[7] Drinkwater, L.E., Wagoner, P., and Sarrantonio, M. Legume-based cropping systems have reduced carbon and nitrogen losses. Nature. 1998;396 262—264.

[8] Frerichs R. Organic Food-grade Corn, 2003 – http://web.aces.uiuc.edu/value/fact-sheets/corn/fact-organic-corn.htm acssed 10.11.2013

[9]   Porter P.M., Huggins D.R., Perillo C.A., Quiring S.R. and Crookston R.K. Organic and other management strategies with two- and four-year crop rotations in Minnesota. Agron. J. 2003;95 233-244.

[10] Delate K, Cambardella CA. Agroecosystem Performance during Transition to Certified Organic Grain Production. Agron J. 2004;96 1288-1298.

[11] Demo Bielek et al. Regulačné technológie v produkčnom procese polnohospodárskych plodín, SPU Nitra, 2000, 667p.

[12] Hiddink GA, van Bruggen AHC., Termorshuizen AJ, Raaijmakers JM, Semenov AV. Effect of organic management of soils on suppressiveness to Gaeumannomyces graminis var. tritici and its antagonist, Pseudomonas fluorescens. Europ. J. Plant Pathol. 2005;113 417-435.

[13] Sobkowitcz P., Tendziagolska E. Competition and productivity in mixture of oats and wheat. J. Agr. And Crop Sci, 2005;191 377-385.

[14] Walsh E. J., Noonan M. G. Agronomic and quality performance of variety mixtures in spring wheat (Triticum aestivum L.) under Irish conditions. Cereal Res. Commun., 1999;26 427-432.

[15] Pests, (http://www.agroatlas.spb.ru/pests/Zabrus_tenebrioides_en.htm), accessed 01.12.2013.

[16] Mortensen D. A., Byastiaans L., Sattin M. The role of ecology in the development of weed management systems: an autlook. Weed Res., 2000;40 49-62.

[17] Bavec F., Bavec M. Effect of maize plant double row spacing on nutrient uptake, leaf area index and yield. Rost. Vyroba, 2001;47 135-140.

[18] Bavec F, Bavec M. Effects of plant population on leaf area index, cob characteristics and grain yield of early maturing maize cultivars (FAO 100-400). Eur. J. Agron., 2002;16 151-159.

[19] Bavec M, Bavec F, Varga B, Kovačević V. Relationship among yield component in winter wheat (Triticum aestivum L.) cultivars affected by seeding rates. Bodenkultur, 2002;53 151-159.

[20] Deria AM, Bell RW, O'Hara GW. Organic wheat production and soil nutrient status in a Mediterranean climatic zone. J. Sust. Agric. 2003;21 21-47.

[21] L-Baeckstrom G., Hanell U., Svensson G. Nitrogen use efficiency in an 11-year study of conventional and organic wheat cultivation. Commun. Soil Sci. Plant Anal., 2006;37 417-449.

[22] Tu C, Ristaino JB, Hu SJ. 2006. Soil microbial biomass and activity in organic tomato farming systems: Effects of organic inputs and straw mulching, Soil Biol. Biochem. 2006;38 247-255.

[23] Bulson HAJ, Snaydon RW, Stopes CE. Effects of plant density on intercropped wheat and field beans in an organic farming system. J. Agric. Sci. 1997;128 59-71, Part1

[24] Bilsborrow P, Cooper J, Tetard-Jones C, Srednicka-Tober D, Baraniski M, Eyre M, Schmidt C, Shotton P, Volakakis N, Cakmak I, Ozturk L, Leifert C, Wilcockson S. The effect of organic and conventional management on the yield and quality of wheat grown in a long-term field trial. Eu. J. Agron., 2013;51 71-80

[25] Barberi P, Cozzani A, Macchia M, Bonari E. Size and composition of the weed seed-bank under different management systems for continuous maize cropping. Weed Res. 1989;38 319-334.

[26] Pimentel D., Hepperly P., Hanson J., Douds D., Seidel R. Environmental, Energetic, and Economic Comparisons of Organic and Conventional Farming Systems, Bio-Science 2005;55 573-582.

[27] Lotter DW, Seidel R, Liebhardt W.2003. The performance of organic and convention-al cropping systems in an extreme climate year. American J. Alt. Agric., 2003;18 146-154

[28] Lockeretz, W., Shearer, G. and Kohl, D. 1981. Organic farming in the Corn Belt. Sci-ence 1981;211 540-547.

[29] Kuepper G. Organic firld corn production. 2002, 16p.

[30] Delate K, van Dee K. Evaluation of Corn, Soybean and Barley Varieties for Certified Organic Production-Crawfordsville Trial, 2000

[31] Sartori L, Basso B, Bertocco M, Oliviero G. Energy use and economic evaluation of a three year crop rotation for conservation and organic farming in NE Italy. Biosyst. Engin. 2005;91 245-256.

[32] Stearns L., David L. W. Northern Plains Organic Crops Marketing Analysis:Wheat, Oats, Sunflower. Agricultural Economics Report No. 293. Department of Agricultural Economics-Agricultural Experiment Station. North Dakota State University. 1993

[33] Sullivan P. Organic Small Grain Production. ATTRA. http://www.attra.ncat.org/attra-pub/smallgrain.html, 2003, accessed 15.10.2013

[34] Bavec M, Narodoslawsky M, Bavec F, Turinek M. Ecological impact of wheat and spelt production under industrial and alternative farming systems. Renew. Agric Food Syst. 2012; 27 242-250.

[35] Williams, J.T., Ed. Cereals and Pseudocereals, Chapman and Hall, London, 1995

[36] Bavec F, Bavec M. Organic Production and Use of alternative Crops, Taylor and Francis; Boca Raton, New York, London, 2006;250 p.

[37] Mujica, A. et al. Quinoa (Chenopodium quinoa Willd.): Ancestral cultivo Andino, ali-mento del presente y futuro, project FAO presentation, Santiago, Chile, 2001;303 pp.

[38] Kaul HP et al. The suitability of amaranth genotypes for grain and fodder use in Central Europe, Die Bodenkultur, 1996;47 173-181.

[39] Aufhamer W, Lee JH, Kúbler E, Kun M, Wagner S. Anbau und Nutzung der Pseudocerealien Buchweizen (Fagopyrum esculentum), Reismelde (Chenopodium quinoa) und (Amaranthus spp.) als Kornefruchtarten. Die Bodenkultur, 1995;2 1-16.

[40] Henderson TL, Johnson BL, Schneiter AA. Row spacing, plant population, and cultivar effects on grain amaranth in the northern Great Plains, Agron. J., 2000;92 329-334.

[41] Teixeira, D.L., Spehar, C.R. and Souza L.A.C. Agronomic characterization of amaranth for cultivation in the Brazilian Savannah, Pesquisa Agropecuaria Brasileira, 2003;38 45-51.

[42] Bavec F, Grobelnik-Mlakar S. Effects of soil and climatic conditions on ermegence of grain amaranths, Eur. J. Agron., 2002;17 93-103.

[43] Jacobsen SE., Jorgensen, I., and Stolen, O. Cultivation of quinoa (Chenopodium quinoa) under temperate climatic conditions in Denmark, J. Agric. Sci., 2004;122 47-52.

[44] Kalinova J, Moudry J. Evaluation of frost resistance in varieties of common buckwheat (Fagopyrum esculentum Moench), Plant Soil and Environ., 2003;49 410-413

[45] Nabwile Omami, E. Response of amaranth to salinity stress. PhD. Thesis, University of Pretoria, 2005; p102, 235 p.

[46] Matuz, Kuepper, G. Organic Field Corn Production. 2002. ATTRA. http://www.attra.org/attra-pub/PDF/fieldcorn.pdf. accessed 20.1.2013

[47] Robinson, R.G. The Buckwheat Crop in Minnesota, Agr. Exp. Sta. Bul., Univ.. Minnesota, St. Paul, 1980;539 p.

[48] Steadman, K.J. Minerals, phytic acid, tannin and rutin in buckwheat seed milling fractions, J. Sci. Food Agric., 2001;81 1094-1100.

[49] Muchova, Z., Cukova, L., and Mucha, R. Seed protein fractions of amaranth (Amaranthus sp.), Rostl. Vyroba, 2000;46 331-336.

[50] Fleming JE, Galwey NW. Quinoa (Chenopodium quinoa), in Underutilized Crops: Cereals and Pseudocereals, Williams, J.T., Ed., Chapman & Hall,London, 1995. p3.-85.

[51] Iqbal Z et al. Allelopathic activity of buckwheat: Isolation and characterization of phenolics, Weed Sci., 2003;51 657-662.

[52] Rasmussen, C., Lagnaoui, A., Esbjerg, P. Advances in the knowledge of quinoa pests, Food Reviews Int., 2003;19 61-75.

[53] Jamriska P. The effect of variety and row spacing on seed yield of amaranth (Amaranthus ssp.), Rostl. Vyroba, 1998;44 71-76.

[54] Zhu, Y.G. et al. Buckwheat (Fagopyrum esculentum Moench) has high capacity to take up phosphorus (P) from a calcium (Ca)-bound source, Plant Soil., 2002;239 1-8.

[55] Myers, R.L. Nitrogen fertilizer effect on grain amaranth, Agron. J., 1998;90 597-602

[56] Grobelnik-Mlakar S. Agronomic characteristics, potential utilisation and quality of grain amaranth Amaranthus cruentus L. Doct. Thesis. Univ. Maribor, 2012.

[57] Grobelnik-Mlakar S, Jakop M, Turinek M; Robačer M, Bavec M, Bavec F. Protein concentration and amino acid composition in grain amaranth (Amaranthus cruentus L.) as affected by sowing date and nitrogen fertilization. African J. Agric. Res. 2012;7 5238-5246.

[58] Erley GSA et al. Yield and nitrogen utilization efficiency of the pseudocereales amaranth, quinoa, and buckwheat under differing nitrogen fertilization, Eur. J. Agron., 2005;22 95-100.

[59] Johnson DL, Ward S. Quinoa, in New Crops, Janick, J. and Simon, J.E.,Eds, Wiley, New York, 1996;222-227.

[60] Belton P, Taylor J, Eds. Pseudocereals and Less Common Cereals: Grain Properties and Utilization Potential, Springer-Verlag, Berlin, 2002;269 pp.

9

Organic Agricultural Practices Among Small Holder Farmers in South Western Nigeria

Sijuwade Adebayo and Idowu O Oladele

1. Introduction

Organic agriculture and biotechnology are two key innovations that are considered to have beneficial impacts on the future sustainability of agriculture (Wheeler, 2005). Conventional farming has played an important role in improving food and fibre productivity to meet human demands but has been largely dependent on intensive inputs of synthetic fertilizers and pesticides (Tu, Louws, Creamer, Mueller, Brownie, Fager, Bell and Shuijin, 2006). Moreover, the conventional intensive agricultural systems have side–effects which compromise food production in terms of quality and safety. Therefore, problems arising from conventional practices have led to the development and promotion of organic farming system that account of the environment and public health as main concerns (Melero, Ruiz Porras, Herencia and Madejon, 2005). Besides, traditional subsistence smallholding farming can no longer meet the needs and expectation of ever-increasing population of Nigeria (Adomi, Monday-Ogbomo and Inoni, 2003). Increasing agricultural productivity, self-sufficiency and poverty alleviation depend on the acceptance and full utilization of modern inputs, as long been recognized and policy formulation and implementation have been done (Aina 2007). The-Research-Extension-Farmers-Linkage-System (REFILS) has been able to ensure some awareness about the use of modern agro-inputs (Oladele, Sakagomi and Kazunobu 2006).

Organic farming represents a deliberate attempt to make the best use of local natural resources and is an environmental friendly system of farming. It relies much on ecosystem management which excludes external input, especially the synthetic ones. Ander son, Jolly and Green (2005) stated that organic farming is a production system that excludes the use of synthetically manufactured fertilizer, pesticides, growth regulators and livestock feed additives. The system relies on crop rotation, crop residues, animal manures, legumes, green manures, off-farm organic wastes, mechanical cultivation and aspects of biological pest control to maintain soil productivity and tilth, to supply plant nutrients and to control insects, weeds and other pests. According to Agbamu (2002), organic farming technology is frequently regarded as the solution to environmental problems that are related to agriculture as well as food safety. Furthermore, Conor (2004) pointed out that organic farming developed as a response to what was perceived to be polluting food supply by modern farming methods and the ensuing degradation of the environment with chemical and other by-products of the industry.

Soil quality is a necessary indicator of sustainability land. The two farming systems (organic and conventional) studied at farm level in Central Italy has emphasized interesting differences on soil quality. It became obvious that organic management affects soil microbiological and chemical properties by increasing soil nutrient availability, microbial biomass and microbial activity, which represent a set of sensitive indicators of soil quality. (Marinari, Mancinelli, Campiglia, Grego, 2006). The bacterial biomass that perform soil functions and resist environmental stress occurring under organic farms scores higher than in other farming systems (Mulder, De Zwart, Van Wijnen, Schouten, Breure, 2003). Furthermore, the results confirm the positive effects of organic manures and diversified crop rotations on soil quality aspects. Rigby and Ca'ceres (2001) and Defoer (2002) reported that organic agriculture tends to conserve soil fertility and system stability better than conventional farming systems. The Food and Agriculture Organization of the United Nations regards organic agriculture as an effective strategy for mitigating climate change and building robust soils that are better adapted to extreme weather conditions associated with climate changes (IFOAM, 2009; Pretty, 1999).

Organic agriculture promotes food safety and quality. The past decade has been characterized by escalating public concern towards nutrition and health and food safety issues (Crutchfield & Roberts, 2000). As a result, at present, consumers perceive relatively high risks associated with the consumption of conventionally grown produce compared with other public health hazards (Williams & Hammitt, 2000, 2001). Mitchell, Hong, Koli, Barrett, Bryant, Denison and Kaffka (2007) discovered that fruits and vegetables produced organically have increased levels of flavonoids which are reported to protect against cardiovascular disease (Hertog and Hollman, 1996) and to a lesser extent, against cancer (Knekt, Kumpulainen, Jarvinen, Rissanen, Heliovaara, Reunanen, Hakulinen and Aromaa, 2002) and other age-related diseases such as dementia (Commenges, Scotet, Renaud, Jacqmin-Gadda, Barberger-Gateau and Dartigues, 2000) whereas the levels of flavonoids did not vary significantly in conventional treatment. Furthermore, Lumpkin (2005), and Zug (2006) noted that the use of chemicals in vegetable production has been identified as a major source of health risk and a cause of extensive environmental damage.

Organic agriculture improves ecological health because farmers maintain nutrient balances in soil through locally available organic materials or recycled farm wastes (Park, Stabler and Jones, 2008; Hynes, 2009). Stolze, Piorr, Harring and Dabbert (2000) and Olsson et al (2001) concluded that nutrient balances on organic farms are often close to zero and that energy efficiency is found to be higher in organic farming than in conventional farming. It also encourages ecosystem service which sustains agricultural productivity and resilience and advocates production intensification through ecosystem management. Fertility management in organic farming relies on a long-term integrated approach rather than the more short-term much targeted solutions common in conventional agriculture (Watson et al., 2002). The practice of organic agriculture has been associated with returns on investment because it offers farmers a much more secure income than when they rely on only one or two inputs (Osborne, 2009; Mcguirk, 1990). Besides, organic farm precludes purchases of organic inputs, loans and thus the profit margin made by farmers increases and farmers are better off financially (Sanchez and Swaiminathan, 2005; Mei, Jewison and Greene, 2006).

Unlike organic agriculture, which emphasizes effective soil management and biodiversity, conventional agriculture (also referred to as intensive agriculture) relies on farming a single crop year after year. To overcome the imbalance imposed upon a conventional farm's ecosystem, harmful agents, such as pesticides and synthetic nitrogen fertilizers are used. Unfortunately, conventional agricultural practices exacerbate rather than alleviate the effects of climate change. The consequence of conventional farming's ecological imbalance is a decline in soil organic matter, soil structure, fertility, microbial and faunal biodiversity. Combine these impacts with the nutrient overload that ultimately ends up in waterways, deforestation, and overgrazing that occurs due to changes in land use, and it's not difficult to see why many are now stating that conventional agriculture represents an unsustainable long-term option.

The description of organic agriculture in the preceding section has led to the generation of research output recommended by Agricultural Knowledge and Information Systems (AKIS) in order to enhance organic agriculture and make it more sustainable and profitable. The information generated on organic agriculture by various AKIS has created the need for vegetable farmers to fill the information needs and bridge the gap in their production activities. The way in which information is sought is information seeking behaviour. The study attempts to analyse the information seeking behaviour and adoption of organic farming practices among vegetable farmers in South Western Nigeria.

2. Organic agriculture

Organic agriculture is a holistic production management system which promotes and enhances agro-ecosystem health, including biodiversity, biological cycles, and soil biological activity. It emphasizes the use of management practices in preference to the use of off-farm inputs, taking into account that regional conditions require locally adapted systems. This is accomplished by using, where possible, agronomic, biological, and mechanical methods, as opposed to using synthetic materials, to fulfill any specific function within the system (FAO, 1999). The FAO/WHO Codex Alimentarius guidelines defined organic agriculture as "a holistic production management [whose] primary goal is to optimize the health and productivity of interdependent communities of soil life, plants, animals and people".

Similarly, the International Federation of Organic Agricultural Movements, with over 750 member organizations in 108 countries, defined it as "a whole system approach based upon sustainable ecosystems, safe food, good nutrition, animal welfare and social justice. Organic production therefore is more than a system of production that includes or excludes certain inputs (IFOAM, 2006; IFOAM, 2002). The aim of organic farming is to create integrated, humane, environmentally and economically viable agriculture systems in which maximum reliance is put on local or on-farm renewable resources, and the management of ecological and biological processes. The use of external inputs, whether inorganic or organic, is reduced as far as possible.

Certified organic food and fiber products are those that have been produced according to documented standards. They are foods that are guaranteed to have been produced and processed in a manner that avoids the use of synthetic fertilizers, pesticides, hormones, genetically modified organisms and irradiation, and which strives to enhance natural biological cycles and to meet minimum animal welfare standards.

"Certified organic agriculture" is defined as a certified system of agricultural production that seeks to promote and enhance ecosystem health while minimizing adverse effects on natural resources. It is seen not just as a modification of existing conventional practices, but as a restructuring of whole farm systems. However, "organic agriculture" is not limited to certified organic farms and products but can include all productive agricultural systems that use sustainable, natural processes, rather than external inputs, to enhance agricultural productivity (Scialabba and Hattam, 2002).

Organic farmers adopt practices to conserve resources, enhance biodiversity, and maintain the ecosystem for sustainable production and can lead to increased food production, in many cases we have seen a doubling of yields, which makes an important contribution to increasing the food security of a region (Park et al, 2008). Therefore, Non-certified organic agriculture' is defined as local, often traditional agriculture that is managed more or less in accordance with the principles of organic agriculture, but is not based on certification, trade and premium prices and it promises an alternative development path in rural areas of low-income countries (Halberg et al., 2006).

The principles of organic agriculture according to IFOAM are principle of Health-Organic agriculture should sustain and enhance the health of soil, plant, animal, human and planet as one indivisible; principle of ecology-organic agriculture should be based on living ecological systems and cycles, work with them, emulate them and help sustain them; principle of Fairness-Organic agriculture should build on relationships that ensure fairness with regard to the common environment and life opportunities and principle of care-organic agriculture should be managed in a precautionary and responsible manner to protect health and the well-being of current and future generations and the environment. Literature suggest that the farm, farmer and institutional factors drive farmers to adopt new technologies (De Francesco, Gatto, Runge and Tretini, 2008; Rehman, Mckemey,Yates, Cooke, Garforth, Tranter, Park and Dorward, 2007; Hattam, 2006). Factors such as the financial and social-economic impacts of new technologies, effects of new technologies on the risk of the farm, available resources and technology transfer programme also have an effect on the decision of the farmer to adopt new technologies.

Organic agriculture is fast emerging as the only sustainable long-term approach to food production. Its emphasis on recycling techniques, biodiversity, low external input and high level output strategies make it an ideal replacement for the petroleum intensive agricultural methods that are currently contributing to global warming (IFOAM, 2008; Swift et al, 2004). There are a number of factors indicating that organic agriculture is far more future proof than conventional agriculture. These include ecosystem services (Pimentel et al, 2005 and Stolze et al, 2000); Ecological health (Backer et al, 2009, D'Agostino and Sovacool, 2011); Soil fertility and system stability (Reddy, 2010, Mader et al, 2002); mitigating climate change (FiBL, 2007, Lee, 2005); food safety and quality (Gallagher et al, 2005, Makatouni, 2002; Magnusson et al, 2001 and Torjusen et al, 2001); return on investment and poverty alleviation (Rigby and Caceres, 2001); consumer preferences (Willer and Youssefi, 2007, Chen, 2007 and Mondelaers et al, 2009); value addition (Ohmart, 2003, Mitchell et al, 2007); market niche (Alroe and Noe, 2008) and indigenous knowledge (Tengo and Belfrage, 2004, IFOAM, 2003).

3. Methodology

The area of study is southwestern Nigeria which comprises of six states namely: Oyo, Osun, Ogun, Ondo, Ekiti and Lagos States. Southwest is situated mainly in the Tropical Rainforest Zone, though with swamp forest in the coastal regions in Lagos, Ogun, Ondo and Delta States. The agricultural sector forms the base of the overall development thrust of the zone. The zone covers an area ranging from swamp forest to western up lands, in between are rain forest and the northern parts of Oyo and Ogun states having derived Guinea savannah vegetation. The areas lie between latitude 5 degrees and 9 degrees North and longitude 2 degrees and 8 degrees East. It is bounded by the Atlantic Ocean in the south, Kwara and Kogi states in the north, Eastern Nigeria in the east and Republic of Benin in the west. It has a land area of about 114,271km square representing 12% of the country's total land areas. The high concentration of agricultural activities justifies the choice of the study area (NARP, 1996).

The research design of the study is descriptive and quantitative which is defined by Bless and Higson-Smith (2000), as a study concerned with the condition that exist, practices that prevail, beliefs and attitudes that are held, processes that are on-going and trends that are developing. The study profile organic farming practices in southwestern Nigeria. The population of the study is the entire population of vegetable farmers in the South Western Nigeria. Cluster sampling technique was adopted for selecting the required sample of urban vegetable producers. From literature and preliminary surveys, vegetable production in urban areas that is market oriented is mostly carried out along perennial sources of water or lowlands. This constrains farmers to clusters around these sources of water. Therefore, cluster sampling is considered appropriate. The sampling technique involves random selection of three states in the southwestern Nigeria which were Oyo, Ogun and Ondo. Three local government areas in the urban were selected from each state to give a total number of nine local government areas used for the study. The choice of these Local government areas is based on the dominance of vegetable producers in the different areas. The three local government areas chosen in Oyo state were Akinyele, Egbeda and Ogbomoso south. The three local government areas chosen in Ogun state were Odeda, Obafemi Owode and Abeokuta north. The three local government areas selected in Ondo state were Akure south, Akure north and Ifedore. A cluster of vegetable producers was selected from each of the local government areas to give total of nine clusters.

Fifty producers were randomly selected from each of the nine clusters to give a total sample size of four hundred and fifty respondents for the study.

Data for this study was generated from primary sources based on the objective of the study. Interview schedule was used to elicit information from the respondents. The questionnaire consisted of 14 organic farming practices in southwestern Nigeria from which the respondents indicated use and non-use. These practices are crop rotation, application of compost, mulching of crops, inter cropping, mixed cropping, crop residues, cover crop, animal manure, organic fertilizer, bio control, natural insect predator. A split half technique was used to determine the reliability coefficient with a reliability coefficient of 0.85. The questionnaire was face validated by panel of experts on agricultural extension, agronomist and organic agricultural researcher. The panel consisted of lecturers in agricultural extension and Agronomy. The study took into account the ethical consideration which was addressed through, voluntary participation. Data were analyzed with the Statistical Package for Social Sciences (SPSS) 18.0 using means and standard deviation.

4. Results and discussions

Table 1 shows a list of 14 organic agriculture practices from which the respondents were asked to indicate their use or otherwise using a 2 ponit scale of Yes (2) and No(1). The actual mean is 1.5 due to the rating scale and a mean of greater than 1.5 denoted a use while a mean less than 1.5 denoted non-use. The mean scores of 11 out of 14 practices were above the actual mean which implies the use of these organic agriculture practices. These technologies are: minimum tillage, crop rotation, sanitation, intercropping, green manure, cover crop, fire, compositing, organic fertilizer, animal manure, and mulching. The results revealed the most prominent organic agriculture practices were minimum tillage (1.81, SD=0.9); crop rotation (1.80, SD=0.7) and mulching (1.79, SD=0.6). With respect to the use of minimum tillage, it is the practice that minimises the disturbance of the soil. The soil is not tilled intensively thereby improving the soil structure. It is a cultivation operation whereby soil is disturbed as little as possible to produce crop. Mulch residue from the previous crop is left on the soil surface which aids in retarding weed growth, conserving moisture, and controlling erosion. Therefore, the practice of minimum tillage is a common operation among the farmers that is usually carried out in order to prepare the soil before planting exercise. Baldwin (2006) noted that many organic farmers typically manage weeds mechanically and, therefore, cannot focus on building soil structure in the same way as conservational tillage practitioners which often relies on herbicides for weed control. Instead, organic farmers use innovative practices such as crop rotations, green manuring, and biological pest control to improve the soil structure and conserve soil organic carbon.

Crop rotation as one of the practices can be attributed to the use of indigenous knowledge, where farmers' belief that soil needs rest and some measure should be put in place to ensure soil maintenance and fertility. One of such measures is bush fallowing whereby a farmland that have been cultivated for some number of years is left uncultivated for few years in order to fallow and regain its lost nutrients. Crop rotation is another measure that is used by the farmers for this purpose. In this case, the farm land is not abandoned but crops that are cultivated on the farm are planted in sequence in order to maintain the soil fertility. Crop rotation is a practice that is as old as farming practice itself. Subba Rao (1999) and Stockdale, et al (2000) observed that crop rotations and varieties are selected to suit local conditions having the potential to sufficiently balance the nitrogen demand of crops. Furthermore, Bending and Lincoln, (1999) in their work among the US farmers noted that organic growers commonly plant rapeseed, mustard, and other brassicas as rotation crops to 'clean up' soil during the winter months. Besides, Crop rotations comprising both grass-clover fields and arable crops have shown to be relatively robust in relation to most problems with weeds, pests and diseases (Dubois et al, 1999).

Mulching ranks highly as a cost-effective means of crop residue usage against soil erosion in annual row-cropping systems on sloping lands; and is at the centre of a resurgent soil conservation ethic in much of North America (Shelton et al., 1995). However, it is not commonly used among the vegetable farmers who reported that mulching is predominantly used by yam producers. The findings of Junge et al, (2009) showed that mulching and cover cropping were mostly regarded as not labour-intensive, highly cost-effective, compatible and easy and cheap to adopt. The farmers had a positive impression of the effectiveness as erosion control measures and also mentioned additional advantages, such as the increased soil fertility from the decomposition of organic material and the release of nutrients however disadvantage of mulching was seen in the amount of grass required, the main material used as mulch in the area.

Other organic agricultural practices used by farmers include practices. Farm sanitation (1.69, SD=0.8), intercropping (1.66, SD=0.2), green manure (1.60, SD=0.9) and cover crop (1.55, SD=0.8). Farm Sanitation is keeping the field clean which help in preventing the growth and multiplication of weed, pest and diseases. The reason may be because farmers are also aware of those things that can prevent them from having good yield or output. Farmers go to farm everyday even after the planting period to weed at interval, remove any form of crop residue or decay of dead animal on their farm that can attract pests and diseases to the crop planted and can cause pollution in the environment. Farmers are aware that if weed are left to grow on their plot, it will compete with the crop planted for the available nutrients and will reduce their yield during harvest. Besides, some weeds affect the crop leaving a residual effect on the crop which can affect the taste or the appearance of the crop. Whenever this happened, the farmer will run at a loss because such crop will not attract buyers and may have to be sold at a ridiculous price.

Baumann et al., (2000) showed that intercropping as a cultural method can be used to suppress weeds and reduces pest population because of the diversity of crops grown. According to Sullivan (2003), if susceptible plants are separated by non-host plants that can act as a physical barrier to the pest, the susceptible plant will suffer less damage. Furthermore, intercropping reduced the nitrate content in the soil profile as intercropping uses soil nutrients more efficiently than sole cropping (Zhang and Long Li, 2003).

Organic agriculture practices	Mean	SD
Minimum tillage	1.81	0.9
Crop rotation	1.80	0.7
Farm Sanitation	1.69	0.8
Intercropping	1.66	0.2
Green manure	1.60	0.9
Cover crop	1.55	0.8
Fire	1.53	0.6
Composting	1.60	0.4
Organic fertilizer	1.68	0.9
Animal manure	1.71	0.3
Mulching	1.79	0.6
Natural pesticides	0.36	0.6
Farm scaping	0.16	0.6
Bio control	0.13	0.3

Table 1. Distribution of the respondents by use of organic agricultural practices

Katyal (2000) reported the application of organic manure as the only option to improve the soil organic carbon for sustenance of soil quality and future agricultural productivity. Wambani et al. (2006) compared the effect of farmyard manure application with recommended rate of inorganic fertilizer and it was discovered that the recommended rate of organic manure was the most profitable and preferred by the farmers because of their low cost, availability of organic manure and longer persistence of kales under these treatments.

Cremer et al; (1996) showed that cover crop residues interfere with the emergence of weed through the allelopathic effect. In addition, Langdale et al. (1991) concluded that cover crops reduced soil erosion by 62 per cent based on a comparison of bare soil and soil planted with a cover crop in the south eastern United States. Results presented for the use of Tithonia and legume cover crops shows increase grain yields significantly in Eastern Uganda (Delve and Jama, 2002). Moreover, Cover crops can improve soil quality (Dabney et al. 2001), and when planted at the beginning of the transition phase, may provide essential soil-building properties and improve weed suppression (Barberi 2002; Martini et al. 2004); however, soil quality effects and ability of cover crops to suppress weed species varies among cover crop species (Melander et al, 2005; Snapp et al, 2005).

The results further shows that the use of organic agricultural practice covered fire (1.53, SD=0.6), composting (1.60, SD=0.4), organic fertilizer (1.68, SD=0.9) and animal manure (1.71, SD=0.3). Wilson (2007) found that flame weeding also called flame cultivation or flaming, is a thermal physical control method that is part of the National Organic Program (NOP) under the organic foods production act of 1990. Flame weeding delays the presence of weeds in crop beds by killing the weeds present before the crop has breached the soil. This can significantly reduce hand-weeding labor costs. Farmers see the use of fire as an easy and faster method of clearing the weeds, trees and bushes particularly at the on-set of planting season when the land is prepared. Besides, some farmers believe that when the land is prepared with fire, the ash of the weeds, trees or residues that were burnt will make the soil to be fertile. Farmers see the use of fire for clearing as cost-effective compared to the use of hired labour. Anon (1999) reported that in Iowa, farmer feedback on flame weeding has been positive however burning as labour-saving tool to clear land and to prevent weed infestation is now being brought into question and many development agencies now advocate no-burning. In the communities, however, it is less a question of burning or no-burning but rather when, where, and how to reduce its negative impact (Aalangdong et al., 1999). Some northern farmers have made a conscious decision to cease bush burning with the aim of regenerating organic material (Millar et al 1996). Singh (2003) noted that organic farmers in India reported the capacity of manure (compost) to fulfil nutrient demand of crops adequately and promote the activity of beneficial macro-and micro-flora in the soil. Also, Ouédraogo et al (2001) showed that farmer was aware of the role of compost in sustaining yield and improving soil quality. However, lack of equipment and adequate organic material for making compost, land tenure and the intensive labour required for making compost are major constraints for the adoption of compost technology. Olayide et al (2011) assessing farm-level limitations and potentials for organic agriculture in northern Nigeria, discovered that the current levels of organic fertilizer use as share of the minimum requirements for take-off for organic agriculture in Nigeria was low despite its potentials.

Vanlauwe, (2004) noted that livestock manure is important in maintaining soil organic matter levels, a critical factor in soil health. Additionally, Omiti et al, (1999) noted that animal manure compost is the most common source of soil amendment in organic agriculture in Nigeria and indeed Africa. Farmers are fully aware of the fertilizing value of animal manure as well as the differences, for example, in nutrient release between the manures as also reported by Dittoh (1999) and Karbo et al. (1999). However, Mafongoya et al (2006) reported that in Africa, though, animal manure is one of the mostly used organic inputs, but as the need for increased agricultural production rises; it has been found to be limited in quality and quantity. Williams (1999) reported similar result among farmers in semi-arid West Africa.

However the use of natural pesticides (0.36, SD=0.6), farm scaping (0.16, SD=0.6) and Bio control (0.13, SD=0.3) were below the actual mean which indicate non-use by the farmers. This may be because these practices do not fit in to the farming system in the study areas. It can also be attributed to the technicality of the use of these practices usch that the application of the practices and the associated legislation and the process of securing permission for the use of these practices.

5. Conclusions

The paper has shown the nature and trend of the use organic agricultural practices among smallholder farmers in South Western Nigeria by highlighting organic agricultural practices that are prominent and those that were less prominent as well as practices that are not in use. Due to the prevailing opportunities and benefits associated with the use of these practices, this paper recommend that farmers should increase their awareness and use of organic agricultural practices.

Author details

Sijuwade Adebayo and Idowu O Oladele*

*Address all correspondence to: oladele20002001@yahoo.com

Department of Agricultural Economics and Extension, North West University Mafikeng Campus, South Africa

References

[1] Aalangdong, O, I., Kombiok, J, M., & Salifu, A, Z., 1999, Assessment of non-burning and organic-manuring practices, ILEIA newsletter, 15(1/2): 47–48

[2] Adomi, E, E; Monday,- Ogbomo, O., & O.E .Inoni, O, E., 2003, Gender factors in crop farmers' access to agricultural information in rural areas of Delta State, Nigeria, *library Review* 52(8):388-93.

[3] Agbamu, J, U., 2002, Agricultural Research Extension Farmer Linkages in Japan: "Policy issues for Sustainable Agricultural Development in Developing Countries" *International Journal of Social and Policy Issues*, 2002(1): 252-263.

[4] Aina, L, O., 2007, Globalization and small-scale farming in Africa: What role for information centers. World Library and Information Congress: 73rd IFLA General Conference and Council, Durban, South Africa, August 19-23, 2007. Accessed February 6, 2010.http://www.ifla.org/iv/ifla73/index.html.

[5] Alrøe, H, F., & Noe, E., 2008, What Makes Organic Agriculture Move: Protest, Meaning or Market? A Polyocular Approach to the Dynamics and Governance of Organic Agriculture, *International Journal of Agricultural Resources, Governance and Ecology*, 7, (1/2), 2008

[6] Anderson, J, B., Jolly, D, A., & Green, R., 2005, Determinants of farmer adoption of organic production methods in the fresh-market produce sector in California: A lo-

gistic regression analysis. A paper presented at the Western Agricultural Economics Association 2005 Annual Meeting, July 6-8, 2005, San Francisco, California. http:// ageconsearch.umn.edu/bitstream/36319/1/sp05an01.pdf, accessed July 2011.

[7] Backer, E, D., Aertsens, J., Vergucht, S., & Steurbaut, W., 2009, Assessing the ecological soundness of organic and conventional agriculture by means of life cycle assessment. *British Food Journal* 111 (10):1028-1061.

[8] Baldwin, K, R., 2006, Organic Production- Conservation of Tillage on Organic Farms, Published by North Carolina Cooperative Extension Service.

[9] Barberi, P., 2002, Weed management in organic agriculture: are we addressing the right issues? *Weed Response* 42:177–193.

[10] Baumann, D, T., Bastiaans, L., & Kropff, M, J., 2000, Competition and Crop Performance in a Leek–Celery Intercropping System, *Crop Science* 41:764–774 (2001).

[11] Bending, G, D., & Lincoln, S, D., 1999, Characterization of volatile sulphur containing compounds produced during decomposition of Brassica juncea tissues in soil, *Soil Biology and Biochemistry*, 31: 695-703

[12] Bless, C., & Higson-Smith, C., 2000, Fundamentals of social research methods: An African Perspective, 3rd Edition, Juta Education (Pty) Ltd, Cape Town, pp.37-42

[13] Chen, M, F., 2007, "Consumer attitudes and purchase intentions in relation to organic foods in Taiwan: moderating effects of food-related personality traits", *Food Quality and Preference*, 18 (7): 1008-21.

[14] Commenges, D., Scotet, V., Renaud, S., Jacqmin-Gadda, H., Barberger-Gateau, P., & Dartigues, J, F., 2000, Intake of flavonoids and risk of dementia, *European Journal of Epidemiology* 2000, *16*, 357-363

[15] Connor, J, O., 2004, Organic Matter ,Bi-monthly Magazine of Irish Organic farmers and a growers association West Cork, waterfall, Beara. 2(14)

[16] Creamer, N, G., Bennett, M, A., Stinner, B, R., Cardina J., & Regnier, E, E., 1996, Mechanisms of weed suppression in cover crop-based production systems, *Horticultural Science*, 31:410-413

[17] Crutchfield, S., Buzby J., Frenzen P., Allshouse J., & Roberts D., 2000, The Economics of Food Safety and International Trade in Food Products, United States Department of Agriculture Economic Research Service 1800 M Street NW Washington, DC, 20036

[18] D'agostino, A, L., & Sovacool, B, K., 2011, Sewing climate-resilient seeds: implementing climate change adaptation best practices in rural Cambodia, Mitigation Adaptive Strategy Global Change DOI 10.1007/s11027-011-9289-7 Springer Science+Business Media B.V. 2011

[19] Dabney, S, M., Delgado, J, A., & Reeves, D, W., 2001, Using winter cover crops to improve soil and water quality, *Communication Soil Science Plant Analysis* 32:1221–1250.

[20] De Francesco, E., Gatto, P., Runge, F., & Trestini, S., 2008, Factors affecting farmers' participation in agri-environmental measures: A Northern Italian perspective, *Journal of Agricultural Economics*, 59: 114- 131.

[21] Defoer, T., 2002, Learning about methodology development for integrated soil fertility management, *Agricultural Systems*, 73: 57–81

[22] Delve, R, J., & Jama, B., 2002, Developing organic resource management options with farmers in eastern Uganda, Proceedings of the 17th World Congress of Soil Science, Bangkok, Thailand, 2002

[23] Dittoh, S., 1999, Sustainable soil fertility management: Lessons from action research.Ileia newsletter 15(1/2): 51–52

[24] Dubois, D., Gunst, L., Fried, P., Stauffer, W., Spiess, E., Mader, P., Alfoldi, T., Fliebbach, A., Frei. R., & Niggli, U., 1999, Dok-Versuch: Ertragsentwicklung und Energieeffizienz. Agrarforschung 6, 71-74

[25] FAO, 1999, Organic Agriculture, Food and Agriculture Organization of the United Nations, Rome, <http://www.fao.org/unfao/bodies/COAG/COAG15/X0075E.htm>. Accessed [26 February 1999]

[26] Gallagher, K., Ooi, P., Mew, T., Borromeo, E., Kenmore, P., & Ketelaar, J, W., 2005, Ecological basis for low-toxicity integrated pest management (IPM) in rice and vegetables, In The pesticide detox (ed. J. Pretty), London,UK: Earthscan pp.116-134, 294pp.

[27] Hazlberg, N., Alrøe, H, F., Knudsen, M, T., & Kristensen, E, S., 2006, Synthesis: prospects for organic agriculture in a global context, CAB International 2006. *Global* Development of Organic Agriculture: Challenges and Prospects pp.343-357

[28] Hattam, C., 2006, Adopting certified organic production: evidence from small-scale avocado producers in Michoacaan, Mexico, Unpublished PhD Thesis, University of Reading.

[29] Hertog, M, G, L., & Hollman, P, C, H., 1996, Potential health effects of the dietary flavonol quercetin, *European Journal of Clinical Nutrition* 1996, *50*, 63- 71

[30] IFOAM, 2002, IFOAM – Norms for Organic Production and Processing, International Federation of Organic Agriculture Movements, Bonn (www.ifoam.org)

[31] IFOAM, 2009, Global Organic Agriculture: Continued Growth. BioFacch World Organic Trade Fair 2009 in Nurenberg, Germany

[32] IFOAM, 2003, Organic and Like-Minded Movement in Africa, International Federation of Organic Agriculture Movements (IFOAM), Bonn, 2003: 102-108

[33] Junge, B., Deji, O., Abaidoo, R., Chikoye, D., & Stahr, K., 2009, Farmers' Adoption of Soil Conservation Technologies: A Case Study from Osun State, Nigeria, *The Journal of Agricultural Education and Extension*, 15:3, 257-274

[34] Karbo, N., Bruce, J., & Otchere, E, O., 1999, The role of livestock in sustaining soil fertility in northern Ghana, ILEIA newsletter 15(1/2): 49–50

[35] Katyal, J. C., 2000, Organic matter maintenance: Mainstay of soil quality, *Journal of the Indian Society of Soil Science*, 2000, 48, 704–716.

[36] Langdale, G. W., Blevins R. L., Karlen D. L Mccool K.K., Nearing M.A, Skidmore E.L., Thomas A.W., Tyler D.D., & Williams J.R. 1991, Cover crop effects on soil erosion by wind and water, In W.L. Hargrove (Ed.), *Cover Crops for Clean Water*, Pp. 15-22, Soil and Water Conservation Society, Ankeny, IA.

[37] Lee, D, R., 2005, Agricultural sustainability and technology adoption: Issues and policies for developing countries, *American Journal of Agricultural Economics* 87 (5): 1325-1334

[38] Lumpkin, H., 2005, Organic Vegetable Production: A Theme for International Agricultural Research. Seminar on production and export of organic fruit and vegetables in Asia, FAO corporate Document Repository, http:www.fao.org/DOCREP/006/AD429E/ad429e13.htm

[39] Mader, P., Fliessbach, A., Dubois, D., Gunst, L., Fried, P., & Niggli, U., 2002, "The ins and outs of organic farming", *Science*, 298 (5600): 1889-90.

[40] Mafongoya, P, L., Bationo, A., Kihara, J., Waswa, B, S., 2006, Appropriate technologies to replenish Soil fertility in Southern Africa, *Nutrient Cycling in Agro ecosystem* 76: 127-151

[41] Magnusson, M, K., Arvola, A., & Koivisto-Hursti, U, K., 2001, Attitudes towards organic foods among Swedish consumers, *British Food Journal*, 103:209– 226

[42] Makatouni, A., 2002, What motivates consumers to buy organic food in the UK?: Results from a qualitative study, *British Food Journal*, 104:345–352.

[43] Marinari, S., Mancinelli, R., Campiglia, E., & Grego, S., 2006, Chemical and biological indicators of soil quality in organic and conventional farming systems in Central Italy, *Ecological Indicators* 6 (2006) 701–711

[44] Martini, E, A., Buyer, J, S., Bryant, D, C., Hartz, T, K., & Denison, R, F., 2004, Yield increases during the organic transition: improving soil quality or increasing experience? *Field Crops Research* 86:255–266.

[45] Mei, Y., Jewison, M., & Reene, C., 2006, Organic products market in China, USADA Foreign Agricultural Service, GAIN Report, CH6405, June

[46] Melander, B., Rasmussen, I, A., & Barberi, P., 2005, Integrating physical and cultural methods of weed control—examples from European research, *Weed Science* 53:369–381

[47] Melero, S., Ruiz Porras, J, C., Herencia, J, F., & Madejon, E., 2005, Chemical and bio-chemical properties in a silty loam soil under conventional and organic management, *Soil & Tillage Research* 90 (2006) 162–170

[48] Millar, D., Ayariga, R., & Anamoh, B., 1996, Grandfather's way of doing: gender relations and the yaba-itgo system in Upper East Region, Ghana. In: Reij C, Scoones I and Toulmin C (eds) Sustaining the soil, Indigenous soil and water conservation in Africa, pp 117-125. London: Earthscan Publications

[49] Mitchell, A, E., Hong, Y, J., Koh, E., Barrett, D, M., Bryant, D, E., Denison, R, F., & Kaffka, S., 2007, Ten-Year Comparison of the Influence of Organic and Conventional Crop Management Practices on the Content of Flavonoids in Tomatoes, *Journal of Agricultural and Food Chemistry* 2007, 55, 6154-6159

[50] Mulder, C, H., De Zwart, D., Van Wijnen, H, J., Schouten, A, J., & Breure, A, M., 2003, Observational and simulated evidence of ecological shifts within the soil nematode community of agro ecosystems under conventional and organic farming, *Functional Ecology* 17 (4): 516-525.

[51] NARP, 1996, Staff Appraisal Report, National Agricultural Research Project: Newman, and Newman, J. (1985). Information work: the new divouris. *British Journals of Sociology*, 36 (4): 497-515.

[52] Ohmart, J, L., 2003, "Direct Marketing with Value-added products (or: "Give me the biggest one of those berry tarts!")", University of California Sustainable Agriculture Research and Education Program.

[53] Oladele, O, I., Jun-Ichi, S., & Kazunobu, T., 2006, Research –extension-farmer-linkage system in South western Nigeria, *Journal of Food, Agriculture & Environment* 4(1): 99-102

[54] Olayide, O, E., Anthony, E, I., Arega, D, A., & Vincent, A., 2011, Assessing Farm-level limitations and Potentials for Organic Agriculture by Agro-ecological Zones and Development Domains in Northern Nigeria of West Africa, *Journal of Human Ecology*, 34(2): 75-85 (2011)

[55] Olsson, P., & Folke, P., 2001, Local ecological knowledge and Institutional dynamics for ecosystem management, Study of Lake Racken water shed, Sweden, *Ecosystems* 4: 85-104

[56] Omiti, J, M., Freeman, H, A., Kaguongo, W., Bett, C., 1999, Soil Fertility Maintenance in Eastern Kenya: Current Practices, Constraints, and Opportunities, CARMASAK Working Paper No. 1. KARI/ICRISAT, Kenya

[57] Osborne, B., 2009, Organic farming, Encarta encyclopaedia

[58] Ouédraogo, E., Mando, A., & Zombré, N, P., 2001, Use of compost to improve soil properties and crop productivity under low input agricultural system in West Africa, *Agriculture, Ecosystems & Environment*, 84 (3): 259-266.

[59] Park, Stabler, Jones, 2008, Evaluating the role of environmental quality in the sustainable rural economic development of England, *Environment, Development And Sustainability* .10, 69-88

[60] Pretty, J., 1999, Can sustainable agriculture feed Africa? New evidence on progress, processes and impacts, *Environment, Development and Sustainability* 1: 253–274.

[61] Reddy, B, S., 2010, Organic Farming: Status, Issues and Prospects-A Review. *Agricultural Economics Research Review* Vol. 23 July-December 2010 pp 343-358

[62] Rehman, T., McKemey, K., Yates, C.M., Cooke, R.J., Garforth, C.J., Tranter, R.B., Park, J.R., and Dorward, P.T 2007. Identifying and understanding factors influencing the uptake of new technologies on dairy farms in SW England using the theory of reasoned action, *Agricultural Systems*, 94: 287- 290.

[63] Rigby, D., & Caceres, D., 2001, "Organic farming and the sustainability of agricultural systems", *Agricultural Systems*, 68 (1): 21-40.

[64] Sanche, P, A., & Swaminathan, M, S., 2005, Hunger in Africa: The link between unhealthy people and unhealthy soils, The lancets 365:442-444

[65] Scialabba, N., 2000, Factors Influencing Organic Agriculture Policies with a focus on Developing Countries, IFOAM 2000 Scientific Conference, Basel, Switzerland, 28-31 August 2000. 13p.

[66] Shelton, D, P., Dickey, E, C., Hachman, S, D., Steven, D., & Fairbanks, K, D., 1995, Corn residue cover on soil surface after planting for various tillage and planting systems, *Journal of Soil Water Conservation* 50, 399–404.

[67] Singh, S., & George, R., 2012, Organic Farming: Awareness and Beliefs of Farmers in Uttarakhand, India, *Journal of Human Ecology*, 37(2): 139-149 (2012)

[68] Snapp, S, S., Swinton, S, M., Labarta, R., Mutch, D., Black, J, R., Leep, R., Nyiraneza, J., & O'Neil, K., 2005, Evaluating cover crops for benefits, costs and performance within cropping system niches, *Agronomy Journal*, 97:322–332.

[69] Stockdale, E., et al., 2000, Agronomic and environmental implications of organic farming systems, *Advanced Agronomy*, 2000, 70, 261–327

[70] Stolze, M., Piorr, A., Ha°Ring, A., & Dabbert, S. 2000, "The environmental impact of organic farming in Europe", Organic Farming in Europe, *Economics and Policy*, 6: 23-86 University of Hohenheim, Hohenheim

[71] Subba Rao, I, V., 1999, Soil and environmental pollution – A threat to sustainable agriculture, *Journal of Indian Society of Soil Science*, 1999, 47, 611–633.

[72] Sullivan, P., 2003, Intercropping principles and production practices. Agronomy systems guide, ATTRA (Appropriate Technology Transfer to Rural Areas), 12 pp (http://www.attra.ncat.org)

[73] Tengo, M., & Belfrage, K., 2004, Local management practices for dealing with change and uncertainty: a cross-scale comparison of cases in Sweden and Tanzania. *Ecology and Society*, 9(3):4, 22p. Available at www.ecologyandsociety.org/vol9/iss3/art4

[74] Torjusen, H., Lieblein, G., Wandel, M., & Francis, C, A., 2001, Food system orientation and quality perception among consumers and producers of organic food in Hedmark County, Norway, *Food Quality and Preference* 12:207–216.

[75] Tu, C., Louws, F, J., Creamer, N, G., Mueller, J, P., Brownie, C., Fager, K., Bell, M., & Shuijin, Hu, 2006, Responses of soil microbial biomass and N availability to transition strategies from conventional to organic farming systems, *Agriculture, Ecosystems and Environment* 113 (2006) 206–215

[76] Vanlauwe, B., 2004, Integrated soil fertility management research at TSBF: the framework, the principles, and their application. In: Bationo, A. (Ed.), Managing Nutrient Cycles to Sustain Soil Fertility in Sub-Saharan Africa, Academy Science Publishers, Nairobi.

[77] Watson, C, A., Younie, D., Stockdale, E.A., Cormack, W, F., 2000, Yields and nutrient balances in stocked and stockless organic rotations in the UK, Aspects *Applied Biology* 62, 261–268.

[78] Wheeler, S., 2005, Factors Influencing Agricultural Professionals' Attitudes Towards Organic Agriculture and Biotechnology

[79] Willer, H. and Youssefi, M. 2007, The World of Organic Agriculture – Statistics and Emerging Trends, International Federation of Organic Agriculture Movements (IFOAM), Germany and Research Institute of Organic Agriculture FiBL, Bonn. 77p.

[80] Williams, T, O., 1999, Factors influencing manure application by farmers in semi-arid west Africa, *Nutrient Cycling in Agro ecosystems* 55: 15–22, 1999.

[81] Williams, P, R., & Hammitt, J, K., 2001, Perceived risks of conventional and organic Produce: Pesticides, Pathogens and Natural toxins, *Risk Analysis* 21 (2): 319-330.

[82] Williams, P, R, D., Hammitt, J, K., 2000, A comparison of organic and conventional fresh produce buyers in Boston Area, *Risk Analysis* 20 (5), 735–746.

[83] Zhang, F., & Long Li, 2003, Using competitive and facilitative interactions in intercropping systems enhances crop productivity and nutrient-use efficiency, *Plant and Soil* 248: 305–312, 2003.

[84] Zug, S., 2006, Monga—seasonal food insecurity in Bangladesh—Bringing the information together, The Journal of Social Studies, 111(July–Sept. 2006)

10

Mixtures of Legumes with Cereals as a Source of Feed for Animals

Mariola Staniak, Jerzy Księżak and
Jolanta Bojarszczuk

1. Introduction

Mixtures of spring cereals with legumes are considered good agricultural practice in many European countries, especially in organic and low-input farming system [1, 2]. Cultivation of mixtures contributes to the complementary use of habitat resources and compensatory growth of individual plant species, causing an increased productivity and greater stability of yield [3, 4]. Moreover, the risk of lodging of legumes is significantly reduced. Mixtures limited the negative effects of excessive share of cereals in crop rotation and they are a good forecrop for the succeeding crops. They have a positive effect on the soil fertility, enriching it with nitrogen through a symbiosis of legumes with nodule bacteria and in organic matter due to the huge amount of crop residue left behind [5]. Legume-cereal mixtures are treated with lower doses of nitrogen fertilizer in comparison with the sole cereal, which is advantageous from an economic point of view. The increase in nitrogen dose usually leads to an increase in the yield of cereal component, while the share of legumes seeds in the yield decreases. Mixed crops can be cultivated on soils poorer by one valuation class than individual species cultivated as sole crops. Yielding of mixtures and crop quality largely depends on the selection of components and their participation. Yield of mixture seeds decrease with increasing percentage of legumes at sowing. Cultivation of mixed crops increases protein content in the seeds of cereal component increases the yield of crude protein in the biomass and increases the content of this component in the yield of the seeds mixture. Such crops are also an effective method of weed infestation control and reduce the spread of diseases and pests, which is very important in organic production system [6, 7, 8].

2. Material and methods

Paper was established on the base of the authors' results and literature. The authors' field experiments were conducted according to organic agriculture rules in Agriculture Experimental Stations of Institute of Soil Science and Plant Cultivation – State Research Institute and individual organic farms in different parts of Poland. The results of research were statistically elaborated. The impact of the examined factors experiments on the determined characteristics were assessed using analysis of variance, the half-intervals of confidence being determined by Tukey's test at the significance level of $\alpha = 0.05$.

2.1. The importance of legume-cereal mixtures and benefits from the cultivation

Mixtures of legumes with cereals may be used in different ways. If they are grown for seeds, they can be used for the production of fodder for monogastric animals (pigs and poultry), because of the increased protein content compared to the grains of sole cereals. In turn, if they are cultivated for green forage, they provide valuable roughage for ruminants. They can also be used for plowing, as green manure.

Mixtures make a better utilization of habitat resources than sole crops. Differentiation in the size and depth of the root systems of cereals and legumes allows them to utilization water and nutrients from different soil layers, the result of which is a compensatory growth and development of plants. The research on comparison of mixtures root systems of wheat and barley with peas sown together in alternate and intersecting rows showed that sowing in alternate rows was the least favorable [9]. Competition between the components of the mixtures can involve the access to light. A higher cereal component often results in limiting growth conditions for the accompanying legume by shading, especially under conditions of increased nitrogen. In legumes, photosynthesis is then limited and nitrogen uptake is reduced [10].

Studies on mixtures of yellow lupine with triticale and oats have shown that a competitive potential of a single legume is larger than a single cereal plant, but because of the larger number of cereals in the mixture, their total pressure on legumes is stronger than the pressure of legumes on cereals [11]. The strength of interspecific competition depends on the severity of intraspecific competition, which is largely related to the participation of the individual components. Mixtures of legumes with cereals create the conditions for the formation of allelopathic interactions that have a significant influence on the subsequent development of stand structure and the share of each component in the creation of seeds yield. Secondary metabolites of root exudates may affect rhisospheric organisms, as well as the neighboring plants [12]. Studies [13] have shown that water solutions of root exudates of seedlings of wheat, triticale and barley (an effect of 5 cereal seedlings on 1 seed of legume), after 4 days, strongly reduced the germination of seeds of pea, vetch, blue and yellow lupine, and after 8 days, exudates of barley and wheat caused the loss of germination of pea seeds (Table 1).

An additional benefit of growing legume-cereal mixtures is their effect on soil fertility and its phytosanitary status. Mixtures mitigate the negative effects associated with consecutive sowing of cereals as they become an element which interrupts the continuity of the crops.

Specifications	Germination after 4 days				Germination after 8 days			
	pea	vetch	yellow lupine	blue lupine	pea	vetch	yellow lupine	blue lupine
Control	96.7 a*	93.7 a	97.7 a	85.7 a	96.7 a	95.7 a	98.5 a	86.5 a
Wheat	58.0 b	57.5 b	48.2 b	48.2 b	Rotted	74.5 b	51.0 b	48.2 b
Triticale	59.5 b	77.0 c	68.7 c	67.5 c	62.5 b	90.5 a	80.2 c	70.0 c
Oat	95.5 a	92.2 a	95.0 a	77.5 a	96.0 a	95.5 a	98.0 a	79.2 ac
Barley	46.0 b	78.5 c	62.0 c	50.5 b	Mildewed	88.0 c	65.2 b	52.2 b

* Number within columns followed by the same letters do not differ significantly

Table 1. Influence of cereal root excretions on germination of legume seeds [own study]

Mixtures provide biodiversity resulting from the different morphological characteristics, physiological and sensitivity of individual components to consecutive sowing, and in the case of cereals, this method allows to avoid the negative consequences of their too frequent sowing in the same field and thus reduce the spread of diseases and pests [14].

In recent times, a lot of attention has been paid to the possibility of nitrate leaching from the soil in the positions of winter crops sown after legumes. Significantly lower losses of nitrogen occur after crops of legume-cereal mixtures. The studies have shown that pea grown as sole crop used nitrogen derived from biological fixation in 70%, while in mixed cultivation with cereals-in 99%, contributing indirectly to potentially smaller losses of this component [15]. Also, lysimetric studies have shown that the level of nitrogen leaching in crops of mixture of peas with barley was reduced compared with the treatments where these plants were grown in sole crop. Therefore, in the areas of protection of drinking water, it is recommended to sow legumes in mixtures with cereals [16, 17].

Pre-crop	Seedsyield		Variability of yields in years (%)
	t·ha⁻¹	%	
Yellow lupine	5.28	100	3.5
Pea + lupine	5.17	98	3.8
Oat + lupine	4.98	94	1.8
Oat + peas	4.94	94	3.6
Oat	4.79	91	4.5
Oat + barley	3.63	69	13.6
Oat + triticale	3.23	61	17.4
Triticale + barley	3.17	60	15.6

Table 2. Yields of winter wheat grown as affected by different pre-crops [21]

Legume-cereal mixtures are a good forecrop for root crops, but especially for cereals. They enrich the soil with organic matter and nutrients. For example, crop residues of mixtures of lupine with triticale (straw, stubble and roots) provide 32 kg of nitrogen and 55 kg of potassium

[18]. The value of the leftover position depends on the selection of components, their participation in the stand, the level of yield and soil conditions. Studies have shown that the yields of winter wheat cultivated after the mixture of pea with spring cereals (wheat, barley) were higher from 5 to 27% than after spring barley, while after a mixture of yellow lupine with triticale-by 31% compared to the yields following the triticale [19, 20]. Cereal species significantly affected the forecrop quality of the mixtures. Mixtures of legumes with wheat, barley and triticale were definitely better forecrops for winter wheat than the same cereals cultivated in sole crop, while a mixture with oats influenced on a small increase in the yield of successive crops compared with sole oats [21, 22]. This is due to the characteristic of oats, which has phytosanitary features and is considered as one of good forecrops for winter cereals. However, cereal mixtures were a worse forecrop (Table 2). The share of components seeds in the mixture also affects the catch crop value of the stand. A larger proportion of legumes in the mixture positively affected the yield of the following plants (winter wheat). This reaction was higher on good soils compared to the poorer ones [22, 23].

2.2. Biological nitrogen fixation

Biological process of atmospheric nitrogen fixation by the bacteria of *Rhizobium* and *Bradyrhizobium* that live in symbiosis with legumes has great significance for agriculture. In the symbiosis process, legumes provide the bacteria with carbohydrates, and in return they receive nitrogen assimilated by them, which they use to produce high-value protein. Nitrogen is used by plants in almost 100%, while in the case of mineral fertilizers, the plants generally use not more than 50% of this element. Cereals growing in the vicinity of legumes use the nitrogen assimilated by nodule bacteria, as it is transferred to the soil in the form of aspartic acid or β-alanine. This phenomenon is particularly important in low-input farming systems, especially in organic agriculture, where the biological fixation is the most important source of nitrogen [24, 25, 26].

The amount of nitrogen fixed by the nodule bacteria in the process of symbiosis depends primarily on the species of legume as a component, its share in the mixture and the level of nitrogen fertilization. The lysimetric studies with using ^{15}N have shown that in vetch sown with oats, 90% of the total nitrogen uptake (about 53 kg ha-1) comes from symbiosis, while oat uses about 28 kg of mineral N, which is one third of the nitrogen taken together by plants in the mixture [27]. The studies under mixed sowings of maize with soybean and oat with vetch have shown that exudates of active root nodules of legumes include NO^{3-}ions, which affect the increase of biomass and nitrogen content of non-legume components of these mixtures. The permeation of nitrate ions took place at night-from late evening to early morning [28]. The complementarity in the use of nitrogen by the mixture components was confirmed by a higher uptake of nitrogen with the yield of mixture seeds of oat with pea compared to sole crops of this species (Table 3). Almost twice as high nitrogen uptake with the yield of mixtures seeds compared to seeds yield of barley was significantly associated with a large proportion of the legume seeds in the sown mixture (70%). The research carried out in different European countries (Denmark, Germany, Great Britain, France and Italy) has shown that the overall N resources were used 30-40% more efficiently by pea-barley intercrops compared to the respective sole crops showing a high degree of complementarity between pea and barley across intercrop designs and very different growing conditions in Europe. As a mean of all site around 20% more efficient soil mineral N uptake was achieved by the intercrops than the sole crops. Soil N uptake by barley in intercrops was associated with an increased reliance of pea on N_2-fixation, rising the percent of total N derived from N_2-fixation [29]. The authors indicate the independent of climatic growing conditions, including biotic and abiotic stresses, across European organic farming systems pea-barley intercropping is a relevant cropping strategy to adapt when trying to optimize N use and thereby N_2-fixation inputs to the cropping system.

Treatment	Dose of nitrogen fertilization (kg·ha⁻¹)			Mean
	0	30	60	
Barley	44 a*	55 a	75 a	58 a
Pea	78 b	86 b	93 b	86 b
Barley+pea	96 c	108 c	103 c	102 c

* Number within columns followed by the same letters do not differ significantly

Table 3. Nitrogen uptake with grain yield [own study based on 30]

The dose of mineral nitrogen had a significant impact on the amount of biological fixation of nitrogen by legumes. Increasing dose of mineral nitrogen caused a decrease in the fixation of atmospheric nitrogen, as it is a well-known tendency that if a plant has the possibility to use soil or fertilizer nitrogen, the assimilation of N_2 decreases. The research carried out in Poland has shown that increasing the level of fertilization of legume-cereal mixtures by mineral nitrogen from 0 to 90 kg ha⁻¹ resulted in a significant reduction of atmospheric nitrogen fixation by legumes. Each 10 kg of the nitrogen applied in a dose of 30 and 60 kg ha⁻¹ in a mixture of pea with wheat or pea with barley caused a reduction in the fixation of this element by about 7-8 kg, while at a dose of 90 kg, the reduction was higher by about 9 kg [31, 32]. Biological nitrogen fixation also depends on the soil conditions (nitrogen content, moisture content, pH) and the severity of disease and pests [33]. A large impact on the amount of symbiotic nitrogen fixation has also soil temperature. The optimal temperature range which allows for the maximum nitrogen fixation is (°C): for big-leaved lupine-25, common vetch-20, faba bean-20, field pea-25, blue lupine-20-30, and for soybean-20-25 [34].

2.3. The selected factors of production

2.3.1. Species and varieties

Yielding of legume-cereal mixtures largely depend on the proper selection of species. Spring barley can be a good component for the mixtures with peas, due to similar habitat require-ments, similar length of growing season and high nutritional value of seeds of such mixture, but also spring triticale and spring wheat, especially on the better soils. Naked oats, however, is a weaker component due to its low yield of grain which is also variable in years [35, 36]. Husked cultivars of oats and barley, however, are more useful to mixture with peas compared to the naked forms of these cereals (Table 4). Vetch yields best in a mixture with wheat, but triticale is also a good component for this legume crop (Figure 1). Other studies show, however, that spring barley is a better component for self-finishing cultivar of vetch compared to oats [37]. Lupines yield good in the mixtures with spring triticale and spring wheat, but poorly in the sowings with oats. In the mixture with oats, the yields of lupine seeds are low and variable in years, and their share in the yield of mixtures usually does not exceed 10%. In addition, yellow lupine proves to be more useful for mixtures with spring cereals than blue lupine [38, 39, 40].

Mixtures composition (sowing: seeds number per m²)	Seeds yield (t·ha⁻¹)		
	mixture	cereal	pea
Barley naked (240) + pea (28)	6.35	5.70	0.65
Barley naked (205) + pea (35)	6.46	5.75	0.71
Barley husked (240) + pea (28)	7.20	6.49	0.71
Barley husked (205) + pea (35)	7.51	6.66	0.85
Oat naked (390) + pea (28)	4.20	3.61	0.59
Oat naked (335) + pea (35)	4.19	3.54	0.65
Oat husked (390) + pea (28)	6.19	5.53	0.66
Oat husked (335) + pea (35)	5.97	5.31	0.67

Table 4. Seeds yield and its components of mixtures of husked and naked spring cereals with pea [41]

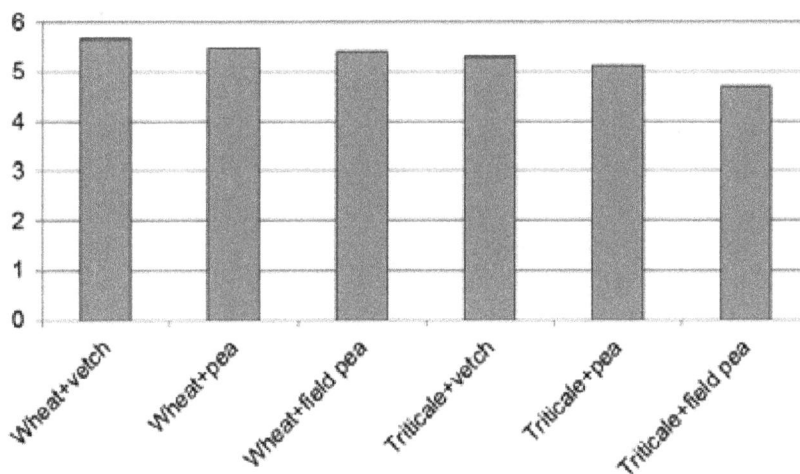

Figure 1. Yields of dry matter depending on the components of mixtures [own study].

On better soils, the choice of cereal species for the mixtures with peas is less important than the choice of legume variety for the cultivation effects [43]. Pea varieties differ considerably in term of the stems length, leaf arrangement, susceptibility to lodging and length of the growing season. They also differ in term of their complementarity in relation to cereals. It is important to choose the cultivars which yield best under given habitat conditions. The height of the components in the mixtures and their diverse habitus determine the canopy architecture. Large differences in the height of plants lead to layered structure of the stand, which creates less favorable light conditions for the species with shorter stems. Particularly unfavorable conditions occur when legumes dominate over cereals, because it leads to the lodging of plants and, consequently, to the yield reduction [44, 45]. High yielding potential was recorded for tendril-leaf cultivars of pea, which are particularly useful for mixtures with spring cereals grown for seeds. Due to the large amount of tendrils, they have a lower coefficient of transpiration and are less susceptible to lodging. Self-finishing cultivars of faba bean are very useful for legume-cereal mixtures, which under favorable soil and moisture conditions yield better in the mixtures with cereals compared to their sole crops [36].

One of the most important factors determining appropriate yielding of mixtures is the share of components seeds at sowing. Cereal sown with legumes gives greater yielding stability of the mixture, but it is also a strong competitor to the legume which causes that the share of legume seeds in the mixture yield is often variable and low. The studies on sowing density and the share of mixture components show that the percentage of legumes seeds should range from 30 to 50%. At tendency to higher lodging, of for example peas, a lower share of about 30% is more favorable [16, 46, 47]. The yields of mixtures primarily depends on the yield of cereal component, while only to a small degree on legume species. Increasing the share of legume seeds in the sowing norm increases their share in the yield, but the yield of grain cereals and the total yield of mixtures generally decrease (Tables 5, 6). This relationship has been confirmed for the mixtures of spring cereals with peas, yellow and blue lupine [39, 40, 42, 45, 47, 48, 49]. The yield and stability of the mixtures are therefore determined by the yield of cereals, and to a lower extent – by legumes. When determining the quantitative composition of the mixtures components of spring cereals with legumes, the following factors should be taken into account: the degree of lodging of the components, plant height, time of maturity and desirable share of legume seeds in the mixture yield, which should range between 20 to 40%.

Lupine share (%)	Cereal species in mixture					
	barley	wheat	triticale	barley	wheat	triticale
	yield of mixture seeds (t·ha⁻¹)			share of lupine seeds (%)		
40	4.03 c*	4.00 c	4.23 b	5.2	7.3	8.0
60	3.74 b	3.80 b	4.17 b	9.4	10.3	10.8
80	3.41 a	3.34 a	3.51 a	14.8	15.4	15.0

* Number within columns followed by the same letters do not differ significantly

Table 5. Yield of seeds mixtures and share of blue lupine seeds in mixtures depending on cereal species and share of blue lupine in sowing [own study].

Mixture components		Sowing rate (number per m²)	Yield					
oat (number per m²)	pea variety		oat seeds		pea seeds		Sum - mixtures	
			t·ha⁻¹	% mean	t·ha⁻¹	% mean	t·ha⁻¹	% mean
	Dawo	30	3.57	118	0.68	73	4.25	108
		45	3.08	102	1.06	115	4.14	105
		60	2.59	86	1.27	137	3.85	98
412	Ramrod	30	3.64	121	0.46	49	4.10	104
		45	3.26	108	0.72	77	3.98	101
		60	2.93	97	1.06	115	3.99	101
	Turkan	30	3.46	115	0.63	68	4.09	104
		45	2.94	98	0.95	103	3.90	99
		60	2.20	73	1.26	136	3.47	88
	Dawo	30	3.38	112	0.81	88	4.19	106
		45	2.90	96	1.07	116	3.97	101
		60	2.43	81	1.52	164	3.95	100
275	Ramrod	30	3.48	116	0.60	65	4.08	104
		45	3.24	108	0.69	75	3.93	100
		60	3.00	100	0.97	105	3.98	101
	Turkan	30	3.30	110	0.63	68	3.93	100
		45	2.74	91	0.95	103	3.69	94
		60	2.06	68	1.34	144	3.39	86

Table 6. Yield of mixtures and their components depending on sowing rate [35]

2.3.2. Soil

The yields of legume-cereal mixtures largely depend on the soil type. On good soils, almost all plant species can be grown and yield very well, because there are no significant restrictions in terms of the selection of individual species. Mixtures may, however, be successfully grown on the worse soils by one quality class than their components grown in sole crops. As a result of differentiated plant growth and development rhythms of individual components, the plants better use of habitat conditions in less favorable soil conditions, as well as in the fields with differentiated soil, deficient water conditions, different forecrops or levels of soil culture [50]. Mixtures of peas with wheat and barley yielded best on Gleyic Phaeozem, Fluvic Cambisol and Haplic Luviosol. The poorest soils are based on sand [51]. On lighter soils (good rye complex), the most efficient mixture was spring triticale with yellow lupine, while slightly worse-triticale with pea and triticale with vetch. Mixtures of oats with yellow lupine and oats with pea also yielded well in such soil conditions [45, 52]. On good soil (good wheat complex), there is no need to cultivate spring cereals with lupines (white, yellow or blue), because the level of their yield has been found not significantly higher than on poorer soils. They are a valuable component, for cultivation on medium and light soils.

An important factor influencing the yield of mixtures is the availability of water in the ground. In the conditions of lower soil moisture, legume-cereal mixtures yielded better than sole crops [53]. Studies have shown that the use of irrigation allows for a cultivation of mixtures for the green matter even on very poor soils, guaranteeing a high level of yield [54].

2.3.3. Fertilization

Nitrogen fertilization is an agrotechnological factor which significantly affects the yield and quality of cereal and legume crops. Legume-cereal mixtures are feeding with lower doses of nitrogen compared to sole cereals. The reason of it is that at lower fertilization, the assimilation of atmospheric nitrogen by legumes increases, and also there crops are less susceptible to the lodging. There is also a smaller competition of cereals in relation to legumes. Increasing nitrogen fertilization results in the dominance of cereal plants, which adversely affects the morphological features and yields of legumes. The reaction of mixtures to the level of nitrogen fertilization also depends on the type of soil and the share of components. A stronger positive effect of fertilization on yield mixtures was observed on lighter soils and at the higher share of cereals [55].

In organic farming, it is necessary to completely resign from mineral nitrogen fertilization, which at the appropriate share of mixture components, good soil and moisture conditions and proper agricultural techniques, does not cause a significant decrease in the yield of mixture seeds. Natural or organic fertilization may be also used, but in a limited extent. Our study showed that in the lack of fertilization, the mixture of barley with peas was the most efficient, but mixtures of oats with peas and oats with vetch provided good yields as well. The significantly weakest was mixture of barley with vetch. In the case of the treatments fertilized with composted manure (at a dose of 30 t ha^{-1}), the mixtures which included peas were more efficient (Table 7). The facts that intercropping of legumes and cereals has produced higher yields than sole cereals without nitrogen fertilization was noticed by several reserchers [56, 57, 58].

Mixture composition	Without fertilization		Fertilization of compost	
	green matter yield (t·ha^{-1})	dry matter yield (t·ha^{-1})	green matter yield (t·ha^{-1})	dry matter yield (t·ha^{-1})
Oat+pea	40.7 b*	9.1 b	39.2 c	8.8 b
Oat+vetch	39.0 b	8.4 b	35.8 b	8.2 b
Barley+pea	41.2 b	10.2 c	38.4 c	8.6 b
Barley+vetch	32.6 a	7.0 a	32.1 a	7.0 a

* Number within columns followed by the same letters do not differ significantly

Table 7. Green and dry matter yields of mixtures cropped on silage [own study]

3. Nutritional value of legume-cereal mixtures (whole-crop silage and grain)

The seeds of legumes are significantly different from the grains of cereals. They contain large amounts of total protein (20 to 40%) and crude fibre and considerably higher amount of minerals compared to the cereals (mainly P), and small amounts of Ca and vitamins. Legume protein is deficient in methionine, a very important amino acid affecting its biological value, but on the other hand, it contains more lysine compared to the cereals. Legume seeds also contain anti-nutritional substances which cause bitter taste of feed and reduce its digestibility and the nutrient availability, so their share in the feed ration should be properly adjusted. It should be noted, however, that as a result of breeding progress there are legume varieties which do not contain or contain only insignificant amounts of anti-nutritional substances, such as alkaloids in lupines or tannins in faba beans and fodder peas, which would limit their feeding.

Cereal species in mixture	Lupine share (%)	Total protein (g·kg⁻¹)	Crude fibre (g·kg⁻¹)	Crude fat (g·kg⁻¹)	Crude ash (g·kg⁻¹)	P (g·kg⁻¹)	K (g·kg⁻¹)
	40	114	26.2	22.2	20.9	3.5	4.4
Barley	60	131	49.3	25.8	24.8	3.8	5.8
	80	139	51.2	27.0	24.6	4.2	5.7
	40	152	33.0	23.0	22.7	3.9	5.8
Wheat	60	177	34.8	26.8	22.4	4.3	5.9
	80	172	40.0	29.4	24.8	4.0	6.4
	40	139	30.6	22.0	24.5	4.2	6.1
Triticale	60	148	39.0	25.1	24.7	4.0	6.2
	80	158	46.0	26.2	27.2	4.6	7.3

Table 8. Content of nutrient components and macroelements in mixture seeds depending on spring cereal species and share of blue lupine [own study]

Growing legume-cereal mixtures significantly enriched feed, especially in protein. High efficiency of protein was recorded in the mixtures of peas with oats, but in terms of the feed quality, mixtures of pea with barley or triticale were favorable as well [59, 60, 61]. The results of the studies on mixtures of peas with barley, oats and wheat have shown that together with the increase of the share of legume, there was also an increase the concentration of protein in cereal grains and its share in the mixture yield [62]. The highest yield of protein was obtained when the share of peas in mixtures with barley and wheat was 75%, and with oats-50%. Our results confirmed these relationships. Increasing the share of blue lupine in two-species mixtures with barley, wheat and triticale resulted in an increase in the total protein content in the yield of mixtures seeds, but the concentration of crude fibre and crude fat has been increased (Table 8). The highest fat and protein content were found in the mixture of lupine with wheat, while the mixture with barley had the lowest amount of these components [40]. In the case of mixtures of peas with barley, there was also an increase in the content of total protein and crude fat in seed yield together with increasing share of legumes at sowing, while the amount of crude fibre decreased (Table 9). The share of pea in the yield had an only insignificant effect on the contents of P, K and Mg [60].

Pea share (%)	Total protein (g·kg⁻¹)	Crude fibre (g·kg⁻¹)	Crude fat (g·kg⁻¹)	Crude ash (g·kg⁻¹)	P (g·kg⁻¹)	K (g·kg⁻¹)	Ca (g·kg⁻¹)	Mg (g·kg⁻¹)
40	183.4	41.0	30.2	25.2	4.4	6.8	0.91	1.3
60	200.6	37.8	31.9	25.6	4.6	8.1	0.92	1.2
80	210.0	35.6	33.7	26.0	4.6	8.6	0.99	1.2

Table 9. Content of nutrient components and macroelements in mixture seeds of pea-barley depending on share of pea [own study]

Legume-cereal mixtures can be grown for green matter and used as a raw material for the production of silage for ruminants or they can be grown for seeds and be used as a component of concentrated feed for monogastric animals. Due to high yields and digestibility of dry matter, the phase of milk-dough stage of cereal is the appropriate term to harvest the mixture for green matter. Silage made from such a mixture may be administered to animals as the only roughage. The results of experiments have shown that silage from the whole-plant legume-cereal mixtures allows to achieve large weight gains of bulls, and the nutrients of that feed are well utilized [63]. It was also found that the energy value of the silage from whole plant of legume-cereal mixtures is similar to maize silage harvested for green matter at milk-dough stage of grain maturity [64]. Dairy cows fed of peas mixed with barley, were characterized a higher milk production compared to the animals fed only with barley. In addition, live weight of cows increased together with the increase in the share of pea in the silage [65]. Better milk production results in cows were obtained when they were fed with silage from the mixtures of pea with barley compared with the mixture of pea with triticale. Higher content of protein and lower content of neutral digestibility fibre (NDF) clearly indicated a more favorable cultivation of such mixture [66]. Protein and energy value of silage from legume-cereal mixtures depends on their composition and the share of individual components. Our study showed that mixtures of pea and vetch with oats had the highest protein and energy value compared to the mixture with barley (Table 10). Taking into account the share of components, it was found that increasing the share of legume seeds at sowing increases digestibility and improves the protein value of the feed made of the mixtures [49].

The nutritional value of legume-cereal mixtures grown for grains mainly depends on the composition of the mixture and the share of components. One of the most important criteria for grain quality evaluation is concenration of crude protein. Analysis of grain quality showed that crude protein concentration in the total intercrops grain yields was significantly higher compared with the sole wheat, but was lower than in sole grain legumes (Table 11). The highest

Mixture composition	PDIF* (g·kg DM)	PDIN (g·kg DM)	PDIE (g·kg DM)	UFL (g·kg DM)	UFV (g·kg DM)	Digestibility of DM (%)
Oat+pea	35.0	98.5	86.8	0.75	0.71	66.2
Oat+vetch	34.7	100.7	87.0	0.76	0.70	66.0
Barley+pea	28.6	86.8	82.7	0.73	0.68	66.2
Barley+vetch	28.8	87.2	83.2	0.74	0.69	66.4

*PDIF – protein digested in the small intestine
PDIN – protein digested in the small intestine supplied by rumen-undegraded dietary protein plus protein digested in the small intestine supplied by microbial protein from rumen-degraded protein
PDIE – protein digested in the small intestine supplied by rumen-undegraded dietary protein plus protein digested in the small intestine supplied by microbial protein from rumen-fermented organic matter
UFL – feed unit for lactation
UFV – feed unit for maintenance and meat production

Table 10. Energy and protein value and digestibility of dry matter of legume-cereal mixtures (50% cereals + 50%legumes) [own study]

crude protein concentration was determined in wheat and vetch intercrop grain yield. The same relationships have been found in the studies on mixtures of wheat and spring barley with peas. In addition, it has been shown that these mixtures contained more methionine, and barley also contained more threonine than cereals grown in sole crops [67].

Treatment	Grain yield (t·ha⁻¹)	Crude protein (g·kg⁻¹ of DM)	Crude protein (kg·ha⁻¹ of grain yield)
Wheat	3.15	113	354
Pea	2.37	230	538
Lupine	1.51	259	383
Bean	1.99	287	571
Vetch	1.88	314	593
Wheat+pea	2.93	130	393
Wheat+lupine	2.77	121	356
Wheat+bean	2.84	135	396
Wheat+vetch	3.34	159	553

Table 11. The grain and crude protein yield of spring wheat and legume grown as sole and dual intercrops [58]

The nutritional value of seeds and grains of legume-cereal mixtures is also analyzed by testing the nutritional value of the protein expressed by the ratio of essential egzogenic amino acids-

EAAI (Essential Amino Acid Index) of Oser's and a rate of limiting amino acid-CS (Chemical Score) of Block and Mitchell's. The results showed that mixtures of winter triticale with vetch had higher EAAI rates than mixtures of winter rye with vetch, and isoleucine was the amino acid which limited nutritional value of the protein. On the other hand, the mixture of spring triticale with field pea characterized by higher rate of protein nutrition value EAAI and high content of lysine, isoleucine and threonine in comparison with a mixture of triticale with pea [68, 69].

4. Weed infestation of legume-cereal mixtures

Control of weed infestation in organic farming involves the use of direct methods, involving interventions into the stand and indirect methods of preventive character, such as proper crop rotation, choice of varieties with greater competitiveness against weeds, proper agronomical practices and the use of undersown crops and mixed sowings [70, 71, 72]. Mixed sowings of legumes with cereals strongly compete with weeds than sole crops, but it is also dependent on the composition of the mixture, the share of components, as well as weather and habitat conditions [58]. Our findings showed that among four mixtures of: oats with peas, oats with vetch, barley with peas and barley with vetch, with 50% share of the components at sowing, the mixture of barley with peas was the most weedy, as evidenced by the largest matter and number of weeds (Table 12). This mixture was also characterized by the largest species diversity of weeds, estimated by Shannon index. The most competitive to weeds was the mixture of oats with vetch. The mixture of barley with vetch had the smallest species diversity of weeds, as was estimated by Simpson's index which indicated a clear dominance of one weed species (Figure 2) [73]. Other studies indicate that among the four mixtures of spring wheat with legumes, such as peas, lupine, vetch and faba bean, the mixture of wheat with vetch limited weed infestation the most, while the least competitive was the mixture with lupine. The highest weed infestation was recorded in the sole lupine and pea [58]. Increasing the share of legume in the mixture caused an increase in weed infestation, which indicates higher competitiveness of cereals than legumes in relation to weeds [73, 74, 75]. Weather conditions have also the significant impact on weed infestation of mixtures. The favorable effect of mixtures on reducing of weed infestation discloses more in the wet years, which favor the development of mixtures and weeds [73, 75].

Mixture composition	Fresh matter of weeds (g·m⁻²)	Dry matter of weeds (g·m⁻²)	Number of weeds (plants·m⁻²)	Number of weeds species
Oat + pea	83.1	17.4	20.7	11
Oat + vetch	53.8	11.9	19.9	10
Barley + pea	355.2	37.8	46.1	16
Barley + vetch	183.7	23.0	42.1	11

Table 12. Weeds mass and number in mixtures depending on share of components [own study]

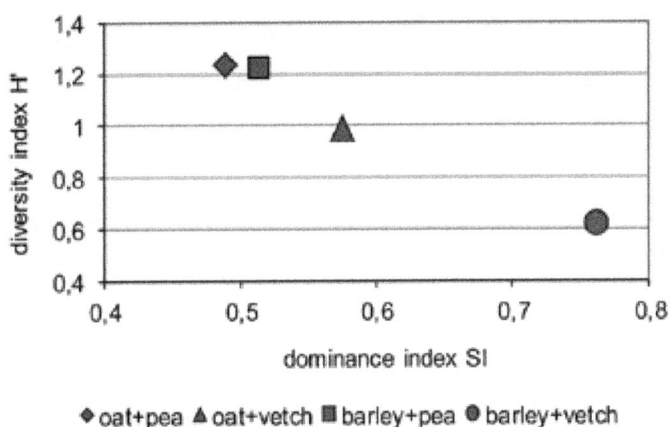

Figure 2. Shannon's diversity index (H') and Simpson's dominance index (SI) for weed flora in mixtures [own study]

5. Undersown crop of serradella in cereals

Intercrop cultivation is one of the agrotechnological ways to reduce adverse changes in agro-ecosystems, which are the result of a large share of cereals in cropping pattern. Their biomass is a significant source of organic matter, and it also has a positive effect on the physical, chemical and biological properties of the soil [76]. Serradella is a species of legume which yields well on poor, slightly acidic soils. Its cultivation provides a number of benefits for animals; provides a valuable, easily digestible feed, positively affects the milk yield of dairy cows, does not contain harmful compounds and it is willingly fed by animals [77]. Undersown into cereal as a support plant, it is more reliable in yielding, less prone to lodging and does not cause trouble during combine harvest. Growing mixed crops also creates more competition for weeds, which allows reducing, and in organic farming, to completely resign from herbicides [72, 78, 79, 80, 81].

Cultivation of serradella has also an ecological importance. The crop residue of this plant contains about 50 kg ha^{-1} of nitrogen, which is largely derived from a biological fixation. Serradella also plays phytosanitary role, reducing the spread of diseases and pests, regenerates the soil and improves balance of organic matter and soil fertility [82, 83]. A limitation in the cultivation of serradella is moisture. The deficiency of water in the initial stages of growth negatively affects the yield of this species [84].

Serradella can be undersown into spring and winter cereals grown for green matter or grains. Our findings showed that undersown serradella positively affected on dry matter yield of spring and winter cereals harvested at the milk-dough stage (for green matter). The highest increase of cereal yield was achieved in the cultivation of serradella with spring barley (Table 13). In the cultivation of cereals for grains (full maturity stage), undersown serradella did not significantly affect the yield of spelt wheat and oats, but limited the yield of spring barley and winter rye compared with the sole crops. The yield of serradella green matter undersown into cereals harvested for green matter was almost two-times higher compared to those harvested in full maturity stage. Regardless of the time of harvesting of the cover crop, the highest yields of serradella green matter were obtained at undersowing into spelt wheat [85, 86].

Specification	Harvest for green matter		Harvest for grain	
	yield of cereal DM (t·ha⁻¹)	green matter of serradella (t·ha⁻¹)	yield of cereal grain (t·ha⁻¹)	green matter of serradella (t·ha⁻¹)
Spelt wheat	4.54		2.61	
Spelt wheat+serradella	5.32	9.26	2.69	5.02
Winter rye	4.97		2.33	
Winter rye+serradella	5.59	7.42	2.06	4.67
Oat	8.36		4.08	
Oat+serradella	8.21	7.34	3.99	4.15
Spring barley	6.27		3.26	
Spring barley+serradella	8.46	8.13	2.54	4.67

Table 13. Yield of cereals in sole crop and intercropped with serradella depending on the harvest time of cover crop [own study]

Serradella undersown into cereals increases the content of total nitrogen and organic carbon in the soil as compared to the contents before sowing of serradella. Furthermore, the amount of nitrogen at harvesting cereals for seeds is higher than at harvesting for green matter. A longer growing period of legumes grown for seeds contributes to a more efficient fixation of atmospheric nitrogen. The process of mineralization of aging roots, nodules as well as leaves, flowers and pods which fell from the lower layers of legumes is started, and nitrogen and other components are released into the soil [86, 87].

6. Conclusions

The cultivation of mixtures of legumes and cereals offers a number of potential agronomic benefits. Coming from two different plant families, legumes and cereals complement each other in the capture of resources. Cereal crops growing in the vicinity of legumes benefit from nitrogen assimilated by legume root nodule bacteria. Increasing the supply of nitrogen by applying fertilizer caused in a substantial reduction of fixation of atmospheric nitrogen by legume crops. Mixtures are particularly relevant to the exploitation of poorer soils which are unsuitable for the production of either component grown as a sole crop. Yielding of the mixtures is highly dependent on the species and proportion of component. The share of legumes in the seed mixture in terms of seed number is recommended to range from 30 to 50%. Total seed yield of mixtures decreases with increasing share of legume seeds in sowing. Increasing the dose of nitrogen for the cultivation of mixtures usually leads to increase in the yield of cereal component, but reduces the proportion of legume seeds in the crop. Legume-cereal mixtures can be grown for green matter and used as a raw material for the production of silage for ruminants or they can be grown for seeds and be used as a component of concentrated feed for monogastric animals. Increasing the share of legume seeds at sowing increases the protein concentration, digestibility and improves the protein value of the feed made of the mixtures.

Legume-cereal mixtures are a good forecrop for cereals. They reduce the negative effects associated with sowing of cereals one after another. Mixtures enrich the soil with organic matter and nutrients, but the value of their post-crop area depends on the choice of components, their share in the stand, the level of yields and soil conditions. Cultivation of cereals after legume-cereal mixtures is characterized by higher yield stability. An ecological importance has also cultivation of serradella as undersown, which plays additionally phytosanitary role, reducing the spread of diseases and pests and regenerates the soil. The benefits of mixed sowings of legumes with cereals are associated with a significant reduction of weed infestation, especially in organic farming. Intercrops are already largely adopted in organic farming, but additional efforts in research are needed for their adoption in more number of farms.

Acknowledgements

The studies have been supported by Ministry of Agriculture and Rural Development of Poland within the multi-annual program of Institute of Soil Science and Plant Cultivation-State Research Institute, task 3.3. Evaluation of efficiency of using different technology elements in the integrated production of basic crops.

Author details

Mariola Staniak, Jerzy Księżak and Jolanta Bojarszczuk

Institute of Soil Science and Plant Cultivation – State Research Institute, Department of Forage Crop Production, Puławy, Poland

References

[1] Watson CA, Atkinson D, Gosling P, Jackson LR, Rayns FW. Managing soil fertility in organic farming system. Soil Use Manage 2002; 18: 239-247.

[2] Knudsen MT, Hauggard-Nielsen H, Jensen ES. Cereal-grain legume intercrops in organic farming – Danish survey. In: European Agriculture in global context: Proceedings of VIII ESA Congress, 11-15 July 2004, Copenhagen, Denmark.

[3] Niggli U, Slab A, Schmid O, Halberg N, Schlüter M. Vision for an Organic Food and Farming Research Agenda to 2025. Report IFOAM EU Group and FiBL 2008.

[4] Doré T, Makowski D, Malézieux E, Munier-Jolain N, Tchamitchian M, Tittonell P. Facing up to the paradigm of ecological intensification in agronomy: Revisiting methods, concepts and knowledge. European Journal of Agronomy 2011; 34(4): 197-210.

[5] Song YN, Zhang FS, Marschner P, Fan FL, Gao HM, Bao XG, Li L. Effect of intercropping on crop yield and chemical and microbiological properties in rhizosphere of wheat (*Triticum aestivum* L.), maize (*Zea mays* L.) and faba bean (*Vicia faba* L.). Biology and Fertility of Soils 2007; 43: 565-574.

[6] Pridham JC, Entz MH. Intercropping spring wheat with cereal grains, legumes, and oilseeds fails to improve productivity under organic management. Agronomy Journal 2008; 100(5): 1436-1442.

[7] Hauggaard-Nielsen H, Jørnsgaard B, Kinane J, Jensen ES. Grain legume–cereal intercropping: the practical application of diversity, competition and facilitation in arable and organic cropping systems. Renewable Agriculture and Food Systems 2008; 23: 3-12.

[8] Corre-Hellou G, Dibet A, Hauggaard-Nielsen H, Crozat Y, Gooding M, Ambus Per Dahlmann C, von Fragstein P, Pristeri A, Monti M, Jensen ES. The competitive ability of pea–barley intercrops against weeds and the interactions with crop productivity and soil N availability. Field Crop Research 2011; 122(3): 264-272.

[9] Tofinga MT, Snaydon RW. The root of cereals and peas when grown in pure stands and mixtures. Plant and Soil 1992; 142: 281-285.

[10] Ofari F, Stern WR. Cereal-legume intercropping systems. Advances in Agronomy 1987; 41: 41-90.

[11] Kotwica K, Rudnicki F. Competition between spring triticale and yellow lupine in mixture. In: Przyrodnicze i produkcyjne aspekty uprawy roślin w mieszankach: Proceedings of the Scientific Conference, 2-3 December 1999, AR Poznań, Poland.

[12] Narwall SS. Allelopathy in crop production. Jodhpur: Scientific Publisher; 1994.

[13] Księżak J, Staniak M. Effect of root excretions from spring cereal seedlings on seed germination of legumes seeds. Journal of Food Agriculture and Environment 2011; 9(3/4): 412-415.

[14] Vasilakoglou IB, Dhima KV, Lithourgidis AS, Eleftherohorinos I. Competitive ability of winter cereal-common vetch intercrops against sterile oat. Experimental Agriculture 2008; 44: 509-520.

[15] Hauggaard-Nielsen H, Ambus P, Jensen ES. The comparison of nitrogen use and leaching in sole cropped versus intercropped pea and barley. Nutrient Cycling in Agroecosystems 2003; 65: 289-300.

[16] Köpke J. Ackerbohnen: eine Gefahr fur das Trinkwasser? Lebendige Erde, Heft 1991; 2(91): 81-87.

[17] Lütke-Entrup N, Groblinghoff F, Stemann, G. Untersuchungen zur Effizienz von Gras-Untersaaten in Ackerbohnen. Gesunde Pflanzen 1993; 45(5): 178-182.

[18] Jasińska Z, Kotecki A. Detailed cultivation of plants. AR Wrocław 1999; cz. II. (Book).

[19] Harasimowicz-Hermann G. Quality of faba bean, cereals and their mixtures as forecrops for winter wheat under at conditions of the Pommerania-Cuiavia region. Zeszyty Problemowe Postępów Nauk Rolniczych 1997; 446: 369–375.

[20] Rudnicki F, Kotwica K. The forecrop value of spring triticale, yellow lupine and their mixture for winter wheat. Fragmenta Agronomica 1994; 2(420): 19-24.

[21] Kotwica K, Rudnicki F. Crop yield of winter wheat after spring triticale and its mixtures with yellow lupine. In: State and perspectives of cereal mixtures cultivation. Poznań AR; 1994. p23-28.

[22] Siuta A, Dworakowski T, Kuźmicki J. Grain yields and forecrop value of mixtures of cereals with leguminous for cereals in conditions of ecological farms. Fragmenta Agronomica 1998; 2(58): 53-62.

[23] Rudnicki F, Kotwica K, Wasilewski P. Forecrop value of cereals-leguminous mixteure for winter cereals. In: Przyrodnicze i produkcyjne aspekty uprawy roślin w mieszankach: Proceedings of the Scientific Conference, 2-3 December 1999, AR Poznań, Poland.

[24] Hauggaard-Nielsen H, Gooding M, Ambus P, Corre-Hellou G, Crozat Y, Dahlmann C, Dibet A, von Fragstein P, Pristeri A, Monti M, Jensen E. S. Pea-barley intercropping for efficient symbiotic N-2-fixation, soil N acquisition and use of other nutrients in European organic cropping systems. Field Crops Research 2009; 113: 64-71.

[25] Bedoussac L, Justes E. Dynamic analysis of competition and complementarity for light and N use to understand the yield and the protein content of a durum wheat–winter pea intercrop. Plant and Soil 2010; 330(1-2): 37-54.

[26] Neumann A., Schmidtke K., Rauber R. Effects of crop density and tillage system on grain yield and N uptake from soil and atmosphere of sole and intercropped pea and oat. Field Crop Research 2007; 100: 285-293.

[27] Triboi E. Détermination in situ de la quantité d'azote fixée symbiotiquement par la vesce en culture associée avec l'avoine. Les Colloques de l'INRA, Paris 1985; 37: 265-270.

[28] Wacquant JP, Ouknider M, Jacquard P. Evidence for a periodic excretion of nitrogen by roots of grass-legume associations. Plant and Soil 1989; 116: 57-68.

[29] Hauggaard-Nielsen H, Gooding M, Ambus P, Corre-Hellou G, Crozat Y, Dahlmann C, Dibet A, Fragstein PV, Pristeri A, Monti M, Jensen ES. Pea–barley intercropping for efficient symbiotic N2-fixation, soil N acquisition and use of other nutrients in European organic cropping systems. Field Crops Research 2009; 113: 64-71.

[30] Podgórska-Lesiak M, Sobkowicz P, Lejman A. Dynamics of nitrogen uptake and utilization efficiency in mixtures of spring barley with field pea. Fragmenta Agronomica 2011; 28(3): 100-111.

[31] Księżak J. Evaluation yields of mixtures pea with spring wheat depending on nitrogen doses. Fragmenta Agronomica 2006; 3: 80-93.

[32] Księżak J. Effect of nitrogen doses on yields of mixtures pea with spring barley. Annales UMCS Lublin 2007; 15: 175-200.

[33] Ledgard SF, Steele KW. Biological nitrogen fixation in mixed legume/grass pastures. Plant and Soil 1992; 141: 137-153.

[34] Liu Y, Wu L, Baddeley JA, Watson CA. Models of biological nitrogen fixation of legumes. A review. Agronomy for Sustainable Development 2010; 31: 155-172.

[35] Rudnicki F, Wenda-Piesik A. Productivity od pea-cereal intercrops on good rye soil complex. Zeszyty Problemowe Postępów Nauk Rolniczych 2007; 516: 181-193.

[36] Księżak J, Borowiecki J. Mixtures of field pea and cereals for fodder production. Proceedings of 4th European Conference on Grain Legumes: Towards the sustainable production of healthy food, feed and novel products. July 8-12, 2001, Cracow, Poland.

[37] Księżak J. Evaluate the usefulness of determinate variety vetch to mixtures with spring cereals. Mat. Sci. Conf.: Plant Breeding, November 19-20, 1997, Poznań, Poland.

[38] Kotwica K, Rudnicki F. Mixture of spring with legumes in the light soil. Zeszyty Problemowe Postępów Nauk Rolniczych 2003; 495: 163-170.

[39] Rudnicki F, Kotwica K. Productivity of lupine-cereal intercrops on very good soil complex). Mat. Sci. Conf.: The economic importance and biology of crop yields mixed, May 11-12, 2006, Poznań, Poland.

[40] Księżak J, Staniak M. Evaluation on mixtures of blue lupine (*Lupinus angustifolius* L.) with spring cereals grown for seeds in organic farming system. Journal of Food Agriculture and Environment 2013; 11(3/4): 1670-1676.

[41] Noworolnik K. Use fullness of naked cultivars of barley and oats to blends with pea. Mat. Sci. Conf.: The economic importance and biology of crop yields mixed, May 11-12, 2006, Poznań, Poland.

[42] Pisulewska E. The effect of botanical composition of spring grain-legume mixtures on protein yield and its amino acid composition. Acta Agraria et Silvestria 1995; ser. Agricultura, 33: 107-115.

[43] Rudnicki F, Wenda-Piesik A. Usefulness estimation of mixtures spring cereals with pea varieties for cropping on good wheat soil complex. Prace Komisji Nauk Rolniczych i Biologicznych 2004; BTN Bydgoszcz, B(52): 309-320.

[44] Księżak J. Diversity of morphological features in selected pea varieties grown in the mixtures with spring barley. Zeszyty Problemowe Postępów Nauk Rolniczych 1998; 463: 389-398.

[45] Rudnicki F, Wenda-Piesik A, Wasilewski P. Sowing of components in pea-barley intercroping on wheat soil complex. Mat. Sci. Conf.: The economic importance and biology of crop yields mixed, May 11-12, 2006, Poznań, Poland.

[46] Krawczyk R, Jakubiak S. Usefulness assessment of selected herbicides in cropping mixture of spring cereals and field pea. Mat. Sci. Conf.: The economic importance and biology of crop yields mixed, May 11-12, 2006, Poznań, Poland.

[47] Księżak J. Cultivation of pea/barley mixtures as a good agricultural practice in production feed for porkers. In: Best practices in agricultural production, Puławy: IUNG; 1998. p263-270.

[48] Droushiotis DN. Mixture of annual legumes and small-grained cereals for forage production under low rainfall. Journal of Agricultural Science (Camb.) 1989; 113: 249-253.

[49] Księżak J, Staniak M. Evaluation of legume-cereal mixtures in organic farming as raw material for silage production. Journal of Research and Applications in Agricultural Engineering 2009; 54(3): 157-163.

[50] Ceglarek F, Buraczyńska D, Płaza A, Bruszewska H. Yielding of mixtures of legumes with spring triticale depending on their composition and term of harvest. In: State and perspectives of cereal mixtures cultivation. Poznań: AR; 1994. p152-156.

[51] Księżak J. The development of pea and spring barley plants in the mixtures on various soil types. Zeszyty Problemowe Postępów Nauk Rolniczych 2007; 516: 83-90.

[52] Wasilewski P. Yielding of cereale-leguminouse mixtures on good rye soil complex. In: Przyrodnicze i produkcyjne aspekty uprawy roślin w mieszankach: Proceedings of the Scientific Conference, 2-3 December 1999, AR Poznań, Poland.

[53] Księżak J. The yield components of mixture of pea with cereals depending on different levels of soil moisture. Roczniki Naukowe AR Poznań 2006; 66: 187-193.

[54] Dudek S, Żarski J. Influence of sprinkling irrigation on yielding of leguminous-cereal crop mixture. Zeszyty Problemowe Postępów Nauk Rolniczych 1997; 442: 389-394.

[55] Noworolnik K. The yields of barley and pea blend depending on nitrogen rate. Mat. Sci. Conf.: The economic importance and biology of crop yields mixed, May 11-12, 2006, Poznań, Poland

[56] Jensen ES. Grain yield, symbiotic N2 fixation and interspecific competition for inorganic N in pea-barley intercrops. Plant and Soil 1996; 182: 25-38.

[57] Lauk R, Lauk E. The yields of legume-cereal mixes in years with high-precipitation vegetation periods. Latvian Journal of Agronomy 2005; 8: 281-285.

[58] Šarūnaitė L, Deveikytė I, Kadžiulienė Ž. Intercropping spring wheat with grain legume for increased production in an organic crop rotation. Žemdirbystė=Agriculture 2010; 97(3): 51-58.

[59] Rudnicki F, Kotwica K. Comparison of the crop effects of the spring cereals-legume mixtures with barley, oats or triticale. Folia Universitatis Agriculturae Steinemsis, Agricultura 2002; 228(91): 125-130.

[60] Staniak M, Księżak J, Bojarszczuk J. Estimation of productivity and nutritive value of pea-barley mixtures in organic farming. Journal of Food Agriculture and Environment 2012; 10(2): 318-323.

[61] Szczygielski T. Yielding of legume-cereal mixtures. Fragmenta Agronomica 1993; 4: 187-188.

[62] Johnston HW, Sanderson JB, Macleod JA. Cropping mixtures of field peas and cereals in Prince Edward Island. Canadian Journal of Plant Science 1978; 58: 421–426.

[63] Wawrzyńczak S, Bielak F, Kraszewski J, Wawrzyński M, Kozłowski J. Maize and grain legume silage for fattening young slaughter cattle. Roczniki Nauk Zootechnicznych 1996; 23(3): 85-97.

[64] Ostrowski R, Daczewska M. The yieding of cereal-pulse mixtures under conditions of the Wielkopolska region, and nutritive value of silages and dried forage for ruminants. Roczniki Nauk Zootechnicznych 1993; 20(2): 157-169.

[65] Skovborg EB, Kristensen VF. Whole-crop barley, peas and field beans for dairy cows. 12. Beretning fra Faellesudvalget for statens planteavls-og husdyrbrugsforøg; 1988.

[66] Blade SF, Lopetinsky KJ, Buss T, Laflamme P. Grain and silage yield of field pea/cereal cropping combinations, In: Abstracts of the 4th European Conference on Grain Legumes. Cracow, 2001.

[67] Pozdísek J, Henriksen B, Løes AK, Ponizil A. Utilizing legume-cereal intercropping for increasing self-sufficiency on organic farms in feed for monogastric animals. Agronomy Research 2011; 9(1/2): 343–356.

[68] Pisulewska E. The effect of botanical composition of spring graini-legume mixtures on protein yield and its amino acid composition. Acta Agraria et Silvestria. Series Agraria. 1995a; 33: 107-115.

[69] Pisulewska E. The total protein content and its amino acid composition of winter grain-legume mixtures as affected by crop production techniques. Acta Agraria et Silvestria. Series Agraria.1995b; 33: 117-126.

[70] Davies DHK, Welsh JP. Weed control in organic cereals and pulses. In: Younie D., Taylor BR., Welsh JP, Wilkinson JM. (eds.) Organic cereals and pulses. Lincoln: Chalcombe Publications; 2001. p77-114.

[71] Höft A, Gerowitt B. Rewarding weeds in arable forming –traits, goals and concepts. Journal of Plant Diseases and Protection 2006; 20: 517-526.

[72] Hauggaard-Nielsen H, Ambus P, Bellostas N, Boisen S, Brisson N, Corre-Hellou G, Crozat Y, Dahlmann C, Dibet A, Fragstein P, Gooding M, Kasyanova E, Launay M, Monti M, Pristeri A, Jensen ES. Intercropping of pea and barley for increased production, weed control, improved product quality and prevention of nitrogen-loses in European organic farming system. Bibliotheca Fragmenta Agronomica 2006; 11(3): 53-60.

[73] Staniak M, Księżak J. Weed infestation of legume-cereal mixtures cultivated in organic farming. Journal of Research and Applications in Agricultural Engineering 2010; 55(4): 121-125.

[74] Buraczyńska D. Weed infestation of legume-cereal mixtures associated with different quantative and qualitive composition. Progress in Plant Protection/Postępy w Ochronie Roślin 2009; 49(2): 779-783.

[75] Bojarszczuk J, Staniak M, Księżak J. Weed infestation of mixture of pea with spring wheat cultivated in organic system. Journal of Research and Applications in Agricultural Engineering 2013; 58(3): 33-40.

[76] Duer I. Effect of catch crops on yield and weed infestation of spring barley. Fragmenta Agronomica 1994; 4(44): 36-45.

[77] Andrzejewska J. Undersown serradella in triticale and winter rye cultivated in monoculture. Part I. Yields of cereals grain and straw. Zeszyty Naukowe ATR Bydgoszcz, Agriculture 1993; 181(33): 61-70.

[78] Hiltbrunner J, Jeanneret P, Liedgens M, Stamp P, Streit B. Response of weed communities to legume living mulches in winter wheat. Journal of Agronomy and Crop Science 2007; 193: 93-102.

[79] O' Donowan JT, Blackshaw RE, Harker KN, Clayton GW, Moyer JR, Dosdall LM, Maurice DC, Tyrkington TK. Integrated approaches to managing weeds in spring-sown crops in western Canada. Crop Protection 2007; 26:390-398.

[80] Staniak M, Bojarszczuk J, Księżak J. Weed infestation of spring cereals cultivated in pure sowing and undersown with serradella (*Ornithopus sativus* L.) in organic farm. Woda-Środowisko-Obszary Wiejskie 2013; 13, 2(42): 121-131.

[81] Bojarszczuk J, Staniak M, Księżak J. The assessment of weed infestation of winter cereals cultivated in pure sowing and undersown with serradella in organic system. Woda-Środowisko-Obszary Wiejskie 2013; 13, 2(42): 5-16.

[82] Wanic M, Majchrzak B, Waleryś Z. Intercrop sowing influence on spring barley yield and balade base diseases at selected lots. Fragmenta Agronomica 2006; 2(90): 149-161.

[83] Płaza A, Ceglarek F. Influence of undercrops on weed infestation of spring barley. Progress in Plant Protection/Postępy w Ochronie Roślin 2008; 48(4): 1463-1465.

[84] Jaskulski D. Effect of companion crops of catch crop on the productivity of crop-rotation link: spring barley – winter wheat. Acta Scientiarum Polonorum. Sec. Agricultura 2004; 3(2): 143-150.

[85] Księżak J, Staniak M, Bojarszczuk J. Evaluation of serradella yielding cultivated as undersown crop in winter cereals harvested at different terms. Journal of Research and Applications in Agricultural Engineering 2013; 58(4): 16-20.

[86] Księżak J, Staniak M, Bojarszczuk J. Yielding evaluation of serradella sown in spring cereals harvested at different times in conditions on the ecological cultivation. Journal of Research and Applications in Agricultural Engineering 2013; 58(4): 12-15.

[87] Harasimowicz-Herman G. Yellow lupine and serradella – the perspective plants in ecological agriculture. Zeszyty Problemowe Postępów Nauk Rolniczych 1997; 446: 307-311.

PERMISSIONS

All chapters in this book were first published by InTech Open; hereby published with permission under the Creative Commons Attribution License or equivalent. Every chapter published in this book has been scrutinized by our experts. Their significance has been extensively debated. The topics covered herein carry significant findings which will fuel the growth of the discipline. They may even be implemented as practical applications or may be referred to as a beginning point for another development.

The contributors of this book come from diverse backgrounds, making this book a truly international effort. This book will bring forth new frontiers with its revolutionizing research information and detailed analysis of the nascent developments around the world.

We would like to thank all the contributing authors for lending their expertise to make the book truly unique. They have played a crucial role in the development of this book. Without their invaluable contributions this book wouldn't have been possible. They have made vital efforts to compile up to date information on the varied aspects of this subject to make this book a valuable addition to the collection of many professionals and students.

This book was conceptualized with the vision of imparting up-to-date information and advanced data in this field. To ensure the same, a matchless editorial board was set up. Every individual on the board went through rigorous rounds of assessment to prove their worth. After which they invested a large part of their time researching and compiling the most relevant data for our readers.

The editorial board has been involved in producing this book since its inception. They have spent rigorous hours researching and exploring the diverse topics which have resulted in the successful publishing of this book. They have passed on their knowledge of decades through this book. To expedite this challenging task, the publisher supported the team at every step. A small team of assistant editors was also appointed to further simplify the editing procedure and attain best results for the readers.

Apart from the editorial board, the designing team has also invested a significant amount of their time in understanding the subject and creating the most relevant covers. They scrutinized every image to scout for the most suitable representation of the subject and create an appropriate cover for the book.

The publishing team has been an ardent support to the editorial, designing and production team. Their endless efforts to recruit the best for this project, has resulted in the accomplishment of this book. They are a veteran in the field of academics and their pool of knowledge is as vast as their experience in printing. Their expertise and guidance has proved useful at every step. Their uncompromising quality standards have made this book an exceptional effort. Their encouragement from time to time has been an inspiration for everyone.

The publisher and the editorial board hope that this book will prove to be a valuable piece of knowledge for researchers, students, practitioners and scholars across the globe.

LIST OF CONTRIBUTORS

Jan Moudrý Jr. and Jan Moudrý
Faculty of Agriculture, University of South Bohemia in České Budějovice, České Budějovice, Czech Republic

Terrence Thomas
North Carolina Agricultural and Technical State University, Department of Agribusiness, Applied Economics and Agriscience Education, USA

Cihat Gunden
Ege University, Faculty of Agriculture, Department of Agricultural Economics, Izmir, Turkey

Maggie Kisaka-Lwayo and Ajuruchukwu Obi
Department of Agricultural Economics & Extension, University of Fort Hare, Alice, South Africa

Sijuwade Adebayo and Idowu O Oladele
Department of Agricultural Economics and Extension, North West University Mafikeng Campus, South Africa

Orhan Özçatalbaş
Akdeniz University, Faculty of Agriculture, Department of Agricultural Economics, Antalya, Turkey

Vytautas Pilipavičius
Aleksandras Stulginskis University, Faculty of Agronomy, Institute of Agroecosystems and Soil Sciences, Akademija, Lithuania

Ilić S. Zoran and Šunić Ljubomir
Faculty of Agriculture Priština-Lešak, Lešak, Serbia

Alvydas Grigaliūnas
The Centre for LEADER Programme and Agricultural Training Methodology, under the Ministry of Agriculture of the Republic of Lithuania, Akademija, Lithuania

Kapoulas Nikolaos
Regional Development Agency of Rodopi, Komotini, Greece

Xiaohou Shao
Key Laboratory of Efficient Irrigation-Drainage and Agricultural Soil-Water Environment in Southern China of Ministry of Education, Hohai University, Nanjing, PR China
Nanjing Ning-ya Environmental Science and Technology Limited Company, Nanjing, PR China

Tingting Chang and Maomao Hou
Key Laboratory of Efficient Irrigation-Drainage and Agricultural Soil-Water Environment in Southern China of Ministry of Education, Hohai University, Nanjing, PR China
College of Water Conservancy and Hydropower Engineering, Hohai University, Nanjing, PR China

Franc Bavec
University of Maribor, Faculty of Agriculture and Life Sciences, Hoce, Maribor, Slovenia

Mariola Staniak, Jerzy Księżak and Jolanta Bojarszczuk
Institute of Soil Science and Plant Cultivation – State Research Institute, Department of Forage Crop Production, Puławy, Poland

Index

www.ingramcontent.com/pod-product-compliance
Lightning Source LLC
Chambersburg PA
CBHW061937190326
41458CB00009B/2760

9 781647 403454